MICROTECHNOLOGY AND MEMS

MICROTECHNOLOGY AND MEMS

Series Editor: H. Baltes   H. Fujita   D. Liepmann

The series Microtechnology and MEMS comprises text books, monographs, and state-of-the-art reports in the very active field of microsystems and microtechnology. Written by leading physicists and engineers, the books describe the basic science, device design, and applications. They will appeal to researchers, engineers, and advanced students.

Mechanical Microsensors
By M. Elwenspoek and R. Wiegerink

CMOS Cantilever Sensor Systems
Atomic Force Microscopy and Gas Sensing Applications
By D. Lange, O. Brand, and H. Baltes

Micromachines as Tools for Nanotechnology
Editor: H. Fujita

Modelling of Microfabrication Systems
By R. Nassar and W. Dai

Laser Diode Microsystems
By H. Zappe

Silicon Microchannel Heat Sinks
Theories and Phenomena
By L. Zhang, K.E. Goodson, and T.W. Kenny

Shape Memory Microactuators
By M. Kohl

Force Sensors for Microelectronic Packaging Applications
By J. Schwizer, M. Mayer and O. Brand

J. Schwizer
M. Mayer
O. Brand

Force Sensors for Microelectronic Packaging Applications

With 149 Figures

 Springer

Dr. Jürg Schwizer
Infineon Technologies
P.O.Box 80 17 09
81609 Munich, Germany
Email: juerg.schwizer@infineon.com

Professor Dr. Michael Mayer
University of Waterloo
Centre for Advanced Materials Joining
E3-2113 Waterloo
Ontario N2L 3G1, Canada
Email: mmayer@mecheng1.uwaterloo.ca

Professor Dr. Oliver Brand
Georgia Institute of Technology
School of Electrical
and Computer Engineering
Atlanta
GA 30332-0250, USA
Email: oliver.brand@ece.gatech.edu

Series Editors:

Professor Dr. H. Baltes
ETH Zürich, Physical Electronics Laboratory
ETH Hoenggerberg, HPT-H6, 8093 Zürich, Switzerland

Professor Dr. Hiroyuki Fujita
University of Tokyo, Institute of Industrial Science
4-6-1 Komaba, Meguro-ku, Tokyo 153-8505, Japan

Professor Dr. Dorian Liepmann
University of California, Department of Bioengineering
466 Evans Hall, #1762, Berkeley, CA 94720-1762, USA

ISSN 1439-6599

ISBN 3-540-22187-5 Springer Berlin Heidelberg New York

Library of Congress Control Number: 2004110615

Springer is a part of Springer Science+Business Media

springeronline.com

© Springer-Verlag Berlin Heidelberg 2005
Printed in Germany

Typesetting by the authors
Cover concept: eStudio Calamar Steinen
Cover production: *design & production* GmbH, Heidelberg
Production: LE-TeX Jeloneck, Schmidt & Vöckler GbR, Leipzig

Printed on acid-free paper   57/3141/YL - 5 4 3 2 1 0

Preface

This book describes a new family of integrated force sensors based on a standard industrial microcircuit technology. The sensors are mainly applied as real-time in situ monitors of the thermosonic wire bonding process commonly used in microelectronics manufacturing.

The background of the book is an eight year long collaboration project between the Physical Electronics Laboratory (PEL) at the Swiss Federal Institute of Technology (ETH), Zurich, and ESEC SA, Cham, both in Switzerland. The work was sponsored by ESEC, the Swiss Commission for Technology and Innovation (CTI), and the Swiss Federal Priority Program MINAST. The project goal was to develop new sensing schemes to better understand bonding processes, facilitate product development and finally improving the existing bonding process, by combining the expertise in integrated sensor development of PEL with the state-of-the-art microelectronic bonding processes of ESEC. During the collaboration project, two of the authors started and finished their disserations which form parts of this book.

In microelectronic manufacturing the thermosonic bonding of gold wires onto aluminum metallization is the most frequently used process for electrical chip to package interconnection. It is a permanent requirement to bond thinner and thinner wires faster and faster. Process mastery is unusually difficult to achieve compared with that of other processes in the field of microelectronics assembly and packaging. We believe miniaturization and cost reduction in this field are tasks that need to be addressed with new technologies, such as the microsensor technology presented in this book.

After an introduction to the wire bonding process in Chap. 1, the reader is guided through the sensor design concept in Chap. 2 and the description of the measurement system in Chap. 3 which also describes an ultrasonic capillary simulation model used to obtain the vibration profile. Chapter 4 summarizes the thorough characterization of the microsensors. Their application for the bonding equipment development, bonding process understanding, and flip-chip reliability characterization is described in Chap. 5, followed by conclusions and an outlook in the last chapter.

The authors are indebted to many colleagues and former students at ESEC and PEL for stimulating discussions, helpful comments, and useful suggestions. In particular, the help of Dr. Daniel Bolliger, Martin Zimmermann, Dr. Christoph Maier, Wan Ho Song, Jan Mattmüller, Antoine Delacrétaz, Stefan Odermatt, Maurice Zaccardi and Andres Erni is gratefully acknowledged. The substantial contributions of Michael Althaus and Quirin Füglistaller to the electronics and software are highly appreciated. With their help the measurement technology presented here has reached the high level of user friendliness which enables any ESEC R&D engineer to use it independently. The continuous support by Prof. Dr. Henry Baltes, director of the Physical Electronics Laboratory (PEL) at ETH Zurich, is gratefully acknowledged. Last but not least our special thanks go to Prof. Dr. Oliver Paul and Hans-Ulrich Müller who, nine years ago, were the first to initiate the real-time use of microsensors for microelectronics packaging processes.

April 2004   *Jürg Schwizer*
Cham, Switzerland, and Atlanta, USA   *Michael Mayer*
                                                              Oliver Brand

Contents

1 Introduction

1.1 Electrical Interconnection Methods

Electrical interconnections are a fundamental part of the semiconductor packaging technology. The trend towards miniaturization, system integration, and manufacturing speed-up is driven by cost optimization of the manufactured product. Reliability is a key issue for packaging technologies. For electrical interconnection, the mature wire bonding technology and the emerging flip-chip techniques are predominant. These techniques establish the electrical contact between chip and substrate through a metal wire or a solder ball, respectively. Figure 1.1 exemplarily shows a schematic cross-section of a wire bonded and a flip-chip bonded package.

The optimization of the packaging process is a pre-condition for high reliability. This is achieved by selecting appropriate materials and process parameters. Improved methods for process monitoring and failure identification are needed to maintain or improve the quality and yield of a packaging process. Current state-of-the-art process characterization methods are often based on off-line tests. These tests are performed after the packaging process and are therefore not real-time measurements. As these tests solely characterize the final state of the packaging process, optimization of the process is generally time-consuming. The origin of the failure mechanisms is difficult to determine as no real-time information is available.

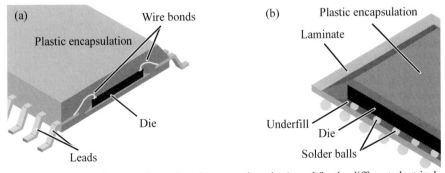

Fig. 1.1. Typical package configurations (cross section view) used for the different electrical connecting methods.

Moreover, the physical quantity causing a device failure may not be accessible to conventional external testing methods. An integration of the probing system into the package thus offers insight into processes that are not accessible to off-line measurement methods. Such systems are often integrated in dedicated dies, denoted packaging test chips. Packaging test chips were demonstrated for examination of reliability issues such as mechanical stresses, corrosion, electrostatic discharge stability, chip surface damages, mobile ions, electromigration, and moisture detection. These test devices are important monitoring tools for examination of mechanical stresses in order to understand the processes taking place during the packaging. An extended overview of the available packaging test chips is found in [1, 2, 3].

Mechanical stress due to mismatch in the thermal expansion coefficients, acting on the interface between the packaging hull and the device is a major source for failure of packaged chips. The study of the stress fields is thus of considerable importance for failure identification and optimization of semiconductor packages. The direct detection of the involved stress fields is done by Moiré interferometry methods [4] or by using stress sensitive devices as e.g. piezoresistors [5, 2], bipolar transistors [6], metal oxide semiconductor field effect transistors (MOSFET) [7], or transverse pseudo-Hall response of MOSFET devices [8]. The subsequent paragraphs summarize some properties of stress sensors based on the piezoresistive effect.

The complete set of the stress fields at distinct places on the chip surface was measured and simulated during thermal cycling of a chip called BMW2 attached to a ceramic substrate with different epoxy adhesives in [9]. Die attachment stress measurements for epoxy substrates are reported in [10]. The stress field components are extracted from resistance changes of piezoresistive serpentine resistors with different orientations integrated on a (111)-oriented silicon substrate as exemplarily shown in Fig. 1.2. The sensor signals are routed to the chip border. Additional diode temperature sensors allow for compensation of the temperature induced resistance change of the stress sensing elements. The test chip BMW2 was also used for recording the surface stress field during the package molding process of the dies [10]. For an eight-element sensor rosette with temperature sensing diode, nine elec-

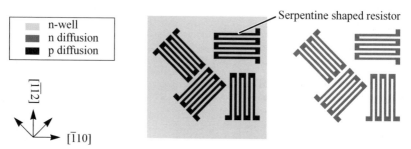

Fig. 1.2. Schematic design of a sensor rosette of p- and n-diffusion piezoresistors used for extraction of the local stress field components from the different resistance changes. Crystal orientation on Si 111 wafer is indicated by arrows.

trical connections have to be routed to the external signal conditioning circuitry. To avoid the resulting large amount of interconnecting lines, on-chip integrated multiplexer circuitry can be used. Such active packaging test chips are reported in [2, 11]. Based on the ATC04 stress test chip family of Sandia National Laboratories, Albuquerque, extensive studies of thermomechanical stresses generated by underfills of flip-chips were performed [11, 12]. The commercially available test chip offers on-chip surface distributed sensor rosettes for measuring stress fields. The piezoresistive sensor elements are connected to a multiplexer circuitry for simplified read-out. All packaging test chips discussed so far are based on rosette-shaped piezoresistive structures to measure the stress field components during various packaging steps.

Investigations of the stress caused by the electro-mechanical contacts during wire bonding, stud bumping or flip-chip bonding are rare in literature. The first real-time application of in situ stress sensors was demonstrated for a ball bond process in [13]. Based on line-shaped piezoresistive sensor structures, made by p-doped diffusions embedded in a n-well of a (100) substrate, the Wheatstone bridge offset change was recorded for various distances between test structure and bonding pad in [14]. Only low-frequency signals are evaluated. No stress measurements at ultrasound frequency are shown. Localized measurements of the stress fields caused by wire bond contacts on passivated surfaces under an applied normal force are presented in [15]. The stress field generated by the bonding tool vibrating at ultrasound frequency is recorded by using line-shaped strain gauges based on on-chip integrated p-doped diffusions [16, 17]. Under the condition of a dominating piezoresistive coefficient Π_{44} for p-diffusions on (100) silicon substrates, the strain values are measured for different distances between the ball contact center and the strain gauge.

So far there were no active test chips available that address the formation and reliability of electro-mechanical contacts such as a wire bonding or flip-chip contact. First steps towards the understanding of the stresses at the contacts are presented in [1, 17, 18]. This book introduces a sensor technology optimized for inspection of the processes that take place during contact formation and reliability tests.

1.2   The Wire Bonding Process

Based on the bonding method, wire bonding machines can be split into two groups, namely ball-wedge bonders and wedge-wedge bonders [19]. Figures 1.3a and (b) show examples of wires bonded with a ball-wedge and a wedge-wedge bonder, respectively. The particularity of a ball-wedge bonder are a different cutting method of the wire, the formation of a ball, and the subsequent formation of the first bond. This entails an arbitrary wire to bonding tool orientation for the ball-wedge bond process with its potential for highest bonding speeds. But it also demands for a large deformation of the wedge in order to cut the wire. This increases the complexity of wedge bond optimization. In this book, experiments are solely performed with a thermosonic ball bonder. However, the use of the sensors is not restricted to the ball bonding method.

(a)

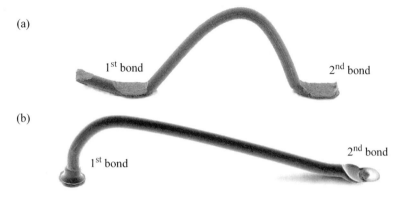

1st bond

2nd bond

(b)

2nd bond

1st bond

Fig. 1.3. Side view of a wedge-wedge bond (**a**) and a ball-wedge bond (**b**). The wire material of the wedge-wedge and ball-wedge bond is AlSi (1 %), and Au (99.99 %), respectively.

In order to establish an electrical connection between die and substrate, two contacts are formed at both ends of a metallic wire (see Figs. 1.1 and 1.3). The contact formation process of a ball bonder differs fundamentally between the first bond on the chip and the second bond on the substrate. Figure 1.4 schematically shows the different bonding stages for the first bond, denoted ball bond, and the second bond, denoted wedge bond. These two bonding steps are part of the wire cycle as shown in Fig. 1.5. The different steps during the wire cycle are explained in the next few paragraphs as the knowledge of the bonding sequence is important for the interpretation of the microsensor measurements that are presented in this book. A reader who is familiar with the wire bonding process may proceed to the next section.

The wire cycle shown in Fig. 1.5 is based on a 80 μm pad-pitch process and has a duration of 92 ms. Shorter wire cycles increase the productivity but may also affect the quality as mechanical vibrations are excited as a result of the high dynamics settings. An optimized design of the bonding head is prerequisite for high speed bonding as described in [20]. In the case of a thermosonic ball bonder, the gold wire is guided inside a ceramic capillary (bonding tool). The tip geometry of the ceramic capillary is important for both the ball and the wedge bond. The geometry is specified by the parameters hole diameter (H), chamfer diameter (CD), face angle (FA), and chamfer radius or chamfer angle (CA). Parameter values for the used capillary SBNE-28ZA-AZM-1/16-XL-50MTA are found in Table 1.1 (see end of this chapter). This capillary suits for bonding processes with 60 μm pad pitch and a 22 μm diameter wire. The definition of the capillary parameters is illustrated in Fig. 1.6.

Prior to the first bond, the gold ball is formed by melting the end of the wire with a spark, also called electrical flame-off (EFO). To form the first contact on the chip, the bond head places the capillary above ball bond position. All bond positions are taught with reference to a coordinate systems based on optical alignment points on the chip and the substrate. Prior to bonding, a camera mounted on the bond heads identifies these alignment points by pattern recognition. The bond head moves

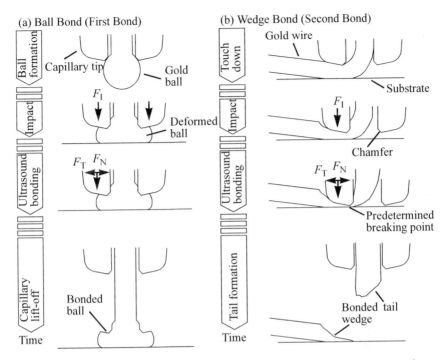

Fig. 1.4. Ball (**a**) and wedge (**b**) bonding sequence. The applied impact force F_I, normal force F_N and tangential ultrasound force F_T are marked with arrows.

down to a security height over the chip as the exact z-position of the chip surface is unknown. From this z-position the capillary moves down with a specific approach speed (A), shown in Fig. 1.5a. This approaching speed will define the impact force (J) to cause the initial deformation of the ball. During this search process, the mechanical vibrations should be small to guarantee a stable impact detection on the force signal. To speed up this process the search height may be reduced under the condition of an optimal z-position estimation of the chip surface, minimized vibrations, and tight path control. After the impact, the normal force is changed to a controlled bond force (K). A feed-back loop controller adjusts the z-position of the bonding head, so that the bond force remains constant. Different approaches may be used to get a signal from the bonding machine that is proportional to the applied bond force. The automatic wire bonder used in this work senses the bond force with a differential pair of piezoelectric cells mounted on the support of the ultrasound transducer system. The measured force signal in Fig. 1.5c contains mechanically caused force oscillations during phases of large acceleration.

During the ultrasound bonding phase, a longitudinal standing wave is excited in a Ti rod, denoted horn. The bonding tool (capillary) is fixed at the end of the horn, shown in Fig. 1.7. Thereby, the longitudinal wave of the rod is transformed into a

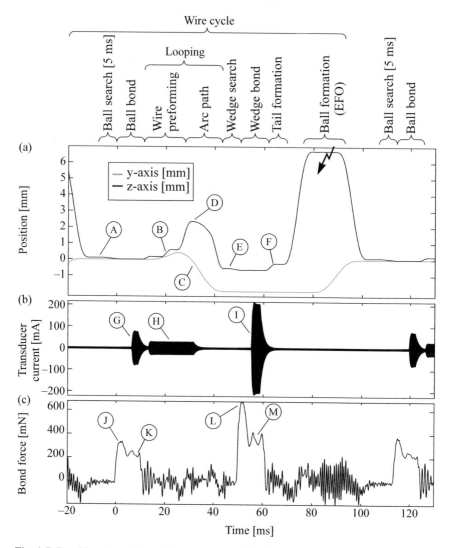

Fig. 1.5. Bond head y- and z-position movement (**a**), ultrasound transducer current (**b**), and bond force (**c**) during a wire cycle measured on a wire bonder 3088iP. The ball position was selected for reference of the axis position. The recorded signals correspond to a 80 μm pad-pitch process. Scaling-down to lower pad-pitches requests lower bond force and ultrasound values. The markers A to M are explained in the text.

transversal oscillation of the ceramic capillary. Capillary shape and material have to be selected according to the bonding application. The capillary tip geometry is of large significance for the wedge bond process. Figure 1.8 illustrates the ultrasonic system and its basic physical quantities. The ultrasound is excited by a stack of

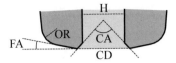

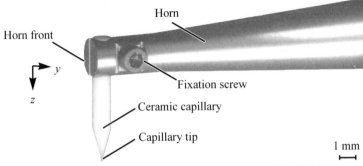

Fig. 1.6. Cross-section of capillary tip. H : hole diameter, CD : chamfer diameter, CA : chamfer angle, FA : face angle, OR : outer radius.

Fig. 1.7. Horn tip with fixed ceramic capillary (bonding tool). Courtesy of ESEC, Cham.

piezoelectric PZT rings. The electronic hardware provides an ac current with the amplitude I_{US} which is phase locked to the horn resonance at frequency f,

$$I(t) = I_{US} \sin(2\pi f t), \tag{1.1}$$

where t is the time, I_{US} is the physical amplitude of the current, and $f = 130$ kHz is the resonance frequency of current ESEC wire bonder transducers. An example of such a current signal is shown in Fig. 1.5b. The piezo stack translates the electrical signals into longitudinal mechanical vibrations along the horn with an amplitude $A_H(t)$ at the tip,

$$A_H(t) = A_H \sin(2\pi f t + \phi_1), \tag{1.2}$$

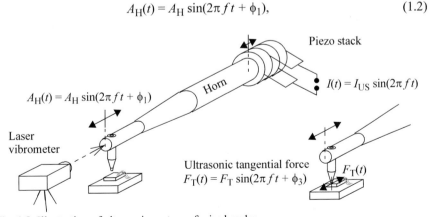

Fig. 1.8. Illustration of ultrasonic system of wire bonder.

where ϕ_1 is a phase difference. The longitudinal vibration translates into a transversal along the capillary with a free air amplitude at the capillary tip $A_C(t)$, defined by

$$A_C(t) = A_C \sin(2\pi f t + \phi_2), \tag{1.3}$$

where the phase difference ϕ_2 is different from ϕ_1 and has been omitted. Once the capillary presses a wire to a bond surface, the process relevant quantity is the ultrasonic force tangential to the wire or chip surface generated by this vibration. This tangential force is

$$F_T(t) = F_T \sin(2\pi f t + \phi_3). \tag{1.4}$$

The transverse ultrasound oscillation, the applied normal force (bond force), and the substrate temperature are the most important machine parameters to cause bond growth for a given substrate type and quality. The substrate temperature is controlled by a heater under the substrate. The substrate quality and uniformity can be improved by plasma cleaning processes prior to bonding [21]. Substrate cleanness is of special importance for the wedge bond process on low-temperature substrates.

After completion of the ultrasound bonding, the capillary is lifted and moved along a distinct path (B) to preform the wire shape while feeding it through the capillary. To reduce friction effects a small ultrasound amplitude can be programmed during the first part of the looping process. The corresponding ultrasound current (H) is shown in Fig. 1.5b. The bond head moves to the bond position of the wedge as indicated in the y-position (C) and z-position. The trajectory from the loop peak height (D) down to the search height of the wedge (E) bond is denoted arc path and is essential for the wire loop shape. Fig. 1.9 shows a schematic view of the different stages of the loop forming process. The bond head movement during the wire pre-

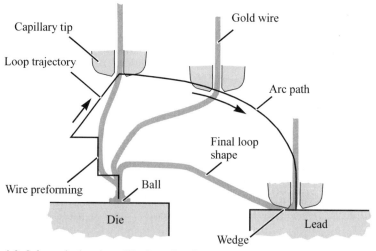

Fig. 1.9. Schematic drawing of the loop forming process.

forming will define the shape and position of the different kinks in the wire. Detailed information on the loop forming process is found in [22]. The process steps of the second bond are shown in Fig 1.4b. The search process for the wedge bond process is equivalent to the search process of the ball. The approaching speed correspondingly defines the wedge impact force (L). Ultrasound oscillation (I) of the capillary may be applied to support bond growth and the formation of a predetermined breaking point. As the ultrasound induces an additional tangential force component, the bond force (M) is often reduced during the ultrasound bonding phase. Special ultrasound and bond force profiles may be used as well. On both sides of the predetermined breaking point, bonds with the substrate are formed, called wedge and tail bond. The tail bond is the bonded area inside the chamfer of the capillary and is needed for the tail forming process. At the end of the bond process the capillary is moved upward by a predefined distance (F). The wire clamp above the capillary is then closed and further movement of the bond head in z-direction will break the wire at the predetermined breaking point. The remaining wire tail that protrudes from the capillary is required to form the gold ball (free air ball) of the subsequent bond. The bond head moves upward to a defined z-position relative to the EFO electrode. A electric discharge between electrode and wire melts the gold at the tip of the tail. This forms the ball for the next ball bond. The arc current and discharge time is employed to control the diameter of the ball. After the EFO process a new wire cycle can be initated. The in the book presented measurements are performed on the ESEC WB 3088iP or the new ESEC WB 3100, shown in Fig. 1.10.

(a)   (b)

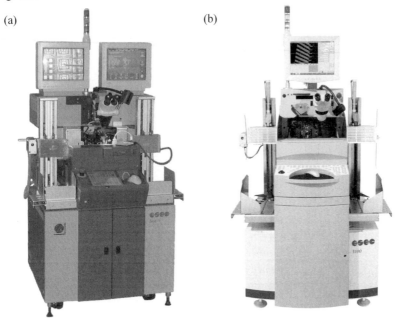

Fig. 1.10. Wire bonder models WB 3088iP (**a**) and WB 3100 (**b**) used for the measurements. Courtesy of ESEC, Cham, Switzerland.

1.3   Measurement Approaches
for Bonding Process Investigation

Ball bond quality inspection is commonly based on a combination of a geometry measurement and bond strength measurement [19, 23]. Ball diameter and ball height contain information about deformation strength and volume of the free air ball. Excessive large or small ball diameters may cause shorts between adjacent balls or unbonded balls (ball non-sticks), and ball height variations will directly affect the loop height stability. Varying ball volumes are an indicator for free air ball or tail formation instabilities. The ultrasound enhanced deformation of the bond can be tracked in situ by observing the z-axis position of the bond head (see Fig. 1.11) under the assumption of a constant free air ball volume. The absolute ball height is not measurable as the exact reference position of the chip surface is unknown and varies from bond position to bond position due to possible chip tilt. The implementation of a feed-back system based on wedge deformation is reported in [24, 25]. The ball geometry provides no reliable information about the actual adhesion at the bonding zone. The shear strength of the contact is widely used to characterize the quality of a bond [19]. Examination of the intermetallic phase distribution of balls, that are released from the pad by a KOH-etching, also reveal bond growth information [26]. As these methods are performed off-line after the bonding process, they only characterize the final stage of the bonding process. To get information about the evolution of the bond growth during ultrasound bonding, an in situ and real-time method is needed. Following approaches are found in literature: PZT sensors in the heater plate [27, 28, 29], PZT sensors mounted on the transducer horn [30], microphones close to the bonding tool [31], laser vibrometer measurements [32], z-position measurement [25], temperature measurements with thermocouple formed with an additional wire [33], temperature measurements with Al-microsensor [34], analysis of ultrasound excitation signals [35], and measurement of flexural waves on the wire [36].

To get a high sensitivity of the measured signal to processes at the contact zone, the distances between contact zone and sensor should be kept as short as possible.

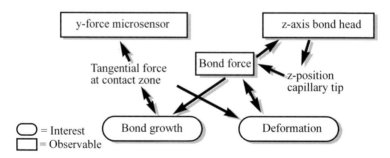

Fig. 1.11. Basic relations between quality parameters of a ball bond (interest) and the signals that are accessible in real-time (observable). Static conditions such as bonding temperature or material quality are not included.

More precise, an ideal sensor picks up the whole physical signal emitted from the contact and suppresses signals from other sources. This book employs the tangential force and normal force at the contact to identify physical processes that happen during the bonding. As the observable processes are dominating the bonding process of Al-Au contacts, information about bond growth can be extracted from the microsensor signals.

Many theories exist about how ultrasound wire bonding can be explained, but a quantitative understanding of the bond process is still missing. One reason for this uncertainty is the lack of suitable measurement methods due to the inaccessibility of the bonding site. An elegant way to overcome this problem is the implementation of the sensing system directly in the semiconductor die that is wire bonded. Such microsensors are advantageous due to their close proximity to the bonding site, offering a high sensitivity to processes taking place during the wire bonding. As the sensing elements are literally fixed to the bonding site, measurements can be performed without any troublesome adjustments. The simple handling significantly speeds up the measurement setup time and furthermore increases the amount of data that can be collected in a certain time. Thus, their application is no more bound to single representative measurements but they can offer information on the whole process window. In contrast to conventional wire bond process characterization methods, such as wire pull and shear test, these integrated microsensors are applicable in situ and real-time. They are therefore highly suitable for studying the time evolution of the bonding process.

Table 1.1. Capillary geometry parameters.

Capillary SBNE-28ZA-AZM-1/16-XL-50MTA	Abbr.	Value
Type		Slimline
Hole diameter	H	28 µm
Chamfer diameter	CD	35 µm
Chamfer angle	CA	90 °
Face angle	FA	11 °
Outer radius	OR	12 µm
Tip diameter	T	80 µm
Bottle neck height	BNH	178 µm
Inner radius	TIR	375 µm
Tool diameter	TD	1585 µm
Bottle neck angle	BNA	10 °
Tool length	TL	11100 µm

2 Sensor Design

This chapter elaborates on the fundamentals of sensors that measure forces at electro-mechanical contacts. Tensor formalism of the piezoresistivity, symmetry considerations, and stress field calculations are the tools for the design process.

2.1 Design Concept

The following design considerations are established for sensors measuring mechanical forces at electro-mechanical contacts for in situ and real-time inspection of the physical processes occurring during or after microelectronic packaging. Mechanical forces at the contacts are modulated or directly caused by nonlinear material behavior that is of considerable relevance in contact formation and reliability.

The piezoresistive effect of monocrystalline silicon is highly qualified for force sensing as the sensor processing seamlessly fits in the standard CMOS process flow. On-chip integrated piezoresistive force sensors can be tailored to access the frequency band from slowly varying thermal expansion forces up to ultrasound frequency oscillations of the bonding tool. By placing the sensor much closer to the electro-mechanical contact than the wavelength of the ultrasonic stress waves, the sensor signals can be interpreted quasi-statically with near-field theory. As the sensor is directly integrated on the die that is subsequently wire or flip-chip bonded, no additional setup-time is needed for alignment of the sensor system to the electro-mechanical contact system. This opens the possibility of real-time measurement of thousands of wires for machine drift and process stability investigations.

High speed operation of wire bonding systems (> 10 wires per second) poses high demands on the performance of the mechanical system. The used accelerations easily result in mechanical cross-coupling between different bonding axes. The detection of the full set of the resulting forces at the bonding site is essential for speed optimization of bonding systems. For instance, if a normal force is applied on the contact zone during the bonding process, bending of the transducer horn can result in an additional force component tangential to the chip surface. This force in y-direction (y-shift force) plays an important role in the wedge bonding process (see *Application*). The ability to measure the full set of forces in real-time plays an important role in the flip-chip applications as well. Consequently, the sensor should enable simultaneous measurement of forces in all three axis directions.

To get a high susceptibility to the actual physical processes, interfering noise sources are suppressed by making use of the symmetry of the local stress fields around the electro-mechanical contact. Moreover, a qualitative understanding of the sensor performance can be derived from symmetry considerations. The international notation [1] is employed for subsequent naming of symmetry operations, as summarized in Table 2.1.

Table 2.1. International notation of symmetry operations.

Symmetry operation	International notation	Specific symmetry operation
Reflection at mirror plane	m	$m_{(100)}$
n-fold rotation	1, 2, 3, 4, 6	$2_{[010]}$
n-fold rotary-reflection	$\bar{1}, \bar{2}, \bar{3}, \bar{4}, \bar{6}$	$\bar{3}_{[111]}$
Inversion	$\bar{1}$	—

Symmetry considerations are highly suited for selecting the spatial arrangement of the sensor elements and qualitative description of their performance. However, when it comes to sensor optimization, the stress fields and the piezoresistive properties have to be known quantitatively. Thus, this chapter also includes a short summary about piezoresistivity and stress field calculation for contact problems. For following considerations Einstein's sum convention is used, i.e.

$$\alpha_{ijkl}\beta_{kl} \equiv \sum_{kl}\alpha_{ijkl}\beta_{kl} \qquad i,j = \{1, 2, 3\}.$$

2.1.1 Piezoresistivity

Piezoresistivity is a change in the electrical resistance caused by a mechanical stress field. In semiconductor resistors, carrier mobility and carrier density modulating effects dominate the piezoresistive behavior compared to pure geometrical effects. Based on the stress-induced redistribution of the conduction/valence band occupancy of the semiconductor, the piezoresistivity pursues the anisotropic energy band structure of the crystal [2]. Consequently, a stress field results in an anisotropic relation

$$E_i = \rho_{ij}J_j \qquad (2.1)$$

between the electrical field strength E_i and the current density vector J_j, connected by the resistivity tensor ρ_{ij}.

To describe the resistance perturbation, the relative change can be expanded in ascending powers of the stress field tensor with piezoresistive coefficients [3].

$$\frac{\Delta \rho_{ij}}{\bar{\rho}} = \pi_{ijkl}^{(1)}\sigma_{kl} + \pi_{ijklmn}^{(2)}\sigma_{kl}\sigma_{mn} + \alpha_1 \Delta T + \ldots \qquad (2.2)$$

The unstressed resistivity $\bar{\rho}$ in equation (2.2) is defined by

$$\bar{\rho} = \frac{1}{3}(\rho_{11} + \rho_{22} + \rho_{33}), \qquad (2.3)$$

an invariant under coordinate transformation. Only the first order piezoresistive term $\pi_{ijkl}^{(1)}$ of equation (2.2) will be included into the sensor design considerations due to the small fraction of the higher order effect in monocrystalline silicon [4]. In addition to stress field induced changes, a temperature change ΔT also affects the resistivity. The constant α_1 is the first order temperature coefficient of the resistivity (TCR). The TCR of highly p-doped (p^+) and n-doped (n^+) silicon is 1600 ppm K^{-1} and 1500 ppm K^{-1} [5], respectively.

The convention for the choice of the coordinate system in relation to the crystal axes is found in [1]. The coordinate axes coincide with the crystal axes [100], [010], and [001] for a cubic system. The piezoresistive coefficients referring to this coordinate system are subsequently marked with a capital Π instead of π. To simplify the notation, the elements of symmetrical matrices are often rearranged by using the reduced index notation, denoted engineering notation, whereby $ij \rightarrow \alpha$ follows the convention $11 \rightarrow 1$, $22 \rightarrow 2$, $33 \rightarrow 3$, $23 \rightarrow 4$, $13 \rightarrow 5$, and $12 \rightarrow 6$. Monocrystalline silicon is a member of the cubic system with class symbol m3m [1]. By using the Neuman's principle, the number of independent piezoresistive coefficients $\Pi_{\alpha\beta}$ degenerates to three independent components Π_{11}, Π_{12}, and Π_{44} [1].

For CMOS processes, the (100) plane of the silicon crystal forms the wafer surface. The primary flat is aligned along the [110] axis. The circuitry and the mechanical dimensions of the chip are aligned along $\langle 110 \rangle$ as shown in Fig. 2.1. The definition of the coordinate system of the wire bonder coincides with [110], [1$\bar{1}$0] and [001] directions. It thus suggests itself to choose a new coordinate system with the axes aligned along the [110], [1$\bar{1}$0], and [001] direction. The transformation from an unprimed to a primed coordinate system is performed with the transformation matrix $a_{ij} = (e'_k)_i (e_k)_j$, defined by the cosine between the new axis vector e'_i and the old axis vector e_j. The transformation of the tensor elements is then given by

$$\begin{aligned} E'_i &= a_{ij}E_j & J'_i &= a_{ij}J_j \\ \rho'_{ij} &= a_{ik}a_{jl}\rho_{kl} & \pi'_{ijkl} &= a_{im}a_{jn}a_{ko}a_{lp}\pi_{mnop}. \end{aligned} \qquad (2.4)$$

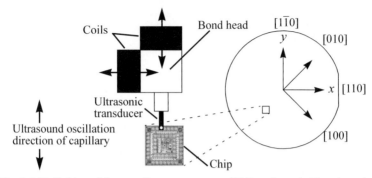

Fig. 2.1. Definition of the coordinate system on a (100) wafer and chip orientation during the wire bonding process.

The tensor transformations for tensors of second order or higher may also be rewritten by the reduced index notation [3]. As the voltage drop of a thin conductor of length L ($L \gg W$) aligned along direction $\boldsymbol{n}$ ($|\boldsymbol{n}| = 1$) is

$$\delta U = E_i n_i \delta L = \rho_{ij} J_j n_i \delta L = \rho_{ij} n_j n_i I \delta L, \tag{2.5}$$

the normalized resistance change of this conductor due to the piezoresistive effect yields to

$$\frac{\Delta R}{R} = \pi_{ijkl} \sigma_{kl} n_i n_j. \tag{2.6}$$

Relation (2.6) is valid for arbitrary coordinate systems with primed piezoresistive tensor, stress field and direction vectors. The primed piezoresistive tensor is related to the unprimed piezoresistive tensor by the transformation rules (2.4).

For subsequent considerations, the stress field components are referred to the coordinate system defined by the axes [110], [1$\bar{1}$0], and [001]. Components related to this coordinate system are indexed with x, y, and z. Equations (2.7) to (2.10) give explicit formulas for the normalized resistance change for resistors aligned along the axes [110], [1$\bar{1}$0], [100], and [010].

[110]

[1$\bar{1}$0]

$$\frac{\Delta R}{R} = \left(\frac{\Pi_{11} + \Pi_{12} + \Pi_{44}}{2}\right)\sigma_{xx} + \left(\frac{\Pi_{11} + \Pi_{12} - \Pi_{44}}{2}\right)\sigma_{yy} + \Pi_{12}\sigma_{zz} \tag{2.7}$$

$$\frac{\Delta R}{R} = \left(\frac{\Pi_{11} + \Pi_{12} - \Pi_{44}}{2}\right)\sigma_{xx} + \left(\frac{\Pi_{11} + \Pi_{12} + \Pi_{44}}{2}\right)\sigma_{yy} + \Pi_{12}\sigma_{zz} \tag{2.8}$$

[010]

$$\frac{\Delta R}{R} = \left(\frac{\Pi_{11} + \Pi_{12}}{2}\right)(\sigma_{xx} + \sigma_{yy}) + (\Pi_{11} - \Pi_{12})\sigma_{xy} + \Pi_{12}\sigma_{zz} \tag{2.9}$$

[100]

$$\frac{\Delta R}{R} = \left(\frac{\Pi_{11} + \Pi_{12}}{2}\right)(\sigma_{xx} + \sigma_{yy}) - (\Pi_{11} - \Pi_{12})\sigma_{xy} + \Pi_{12}\sigma_{zz}. \qquad (2.10)$$

As sensor elements in the (100) plane and the piezoresistive tensor are invariant under symmetry operation $m_{(001)}$, the sensor can not be sensitive to linear terms of σ_{xz} and σ_{yz}. As a consequence, in-plane shear forces at the contact zone can only be measured indirectly. Measuring in-plane shear forces directly would need a material without mirror symmetry $m_{(surface)}$ or integration of vertical sensor elements. The former is the case for (111) wafer substrates [3]. To achieve universal usability, the CMOS mainstream substrates is given priority for a wire bonder and flip-chip test chip application.

The physical origin of the piezoresistive effect results in considerable doping and temperature dependence of the piezoresistive coefficients, described in [6, 7]. The temperature coefficient (TC) and the piezoresistive coefficient both decrease with increasing doping concentration. Due to the implantation process, the doping concentration varies with vertical dimension. To account for the change in impurity density, mobility, and piezoresistive effect, average piezoresistive constants can be defined. From measurement and numerical calculations for typical doping profiles it is concluded, that the piezoresistive coefficients depend mainly on the surface concentration [6].

The presented sensors use highly doped p^+ and n^+ source/drain implantations of the CMOS process for the active piezoresistive sensor area. Table 2.2 lists pub-

Table 2.2. Piezoresistive coefficients of highly doped implantations at 25 °C.

	N_s [cm^{-3}]	Π_{11} [10^{-11}Pa^{-1}]	Π_{12} [10^{-11}Pa^{-1}]	Π_{44} [10^{-11}Pa^{-1}]	Ref.
p^+	$2 \cdot 10^{20}$	—	—	62	[6]
Boron	(a)	—	—	57.2	[8] [b]
	(a)	$\Pi_{11} + \Pi_{12}$: 1.24 ± 1.8		69.0 ± 1	[9] [c]
	(a)	1.1 ± 1.3	1.0 ± 0.6	75.2 ± 2	[10] [c]
n^+	$2 \cdot 10^{20}$	-34	—	-15	[6]
Phosphorous	$2 \cdot 10^{20}$	-28	—	—	[11]
	(a)	—	—	-9.7	[8] [b]
	$1 \cdot 10^{20}$	-30	—	—	[12]
	(a)	$\Pi_{11} + \Pi_{12}$: -16.65 ± 1.5		-14.2 ± 0.5	[9] [c]
	(a)	-28.7 ± 1.3	12.6 ± 1.4	-15.5 ± 1.0	[10] [c]

(a) No surface doping concentration available.
(b) 0.8 μm single poly, double metal CMOS process of Micronas GmbH, Freiburg, Germany.
(c) 0.6 μm CMOS process of TSMC (Taiwan Semiconductor Manufacturing Co.).

lished values of piezoresistive coefficients for highly doped silicon implantations. The specification of the doping surface concentration N_s is essential due to the large gradient of the piezoresistive coefficients for these high doping concentrations.

Wheatstone Bridge Configuration. The application of a Wheatstone bridge configuration offers an advantage over single resistor element transducers as it accounts for the symmetry of the problem and enables the compensation of undesired detrimental thermal and mechanical effects. An overview of the different Wheatstone bridge configurations is found in [13]. Figure 2.2 displays a full Wheatstone bridge configuration and the numbering convention for the resistors. The voltage difference between the sensing terminals U_a and U_b

$$\Delta U = \left(\frac{R_2}{R_1 + R_2} - \frac{R_4}{R_3 + R_4} \right) U \tag{2.11}$$

is a function of the applied supply voltage U and the resistor values. Under the condition of equal nominal resistor values and pertubation of the resistors R_1 and R_4 with opposite sign to resistors R_2 and R_3

$$R_1 = R - \Delta R \qquad\qquad R_2 = R + \Delta R$$
$$R_3 = R + \Delta R \qquad\qquad R_4 = R - \Delta R, \tag{2.12}$$

the normalized output voltage of the bridge simplifies to

$$\frac{\Delta U}{U} = \frac{\Delta R}{R}, \tag{2.13}$$

whereas ΔR is the change in the resistance.

2.1.2   Symmetry Considerations

Making use of the symmetries of the problem helps to cancel parasitic effects and therefore increases the repeatability of measurements. Subsequent considerations suppose a contact between a body with rotational symmetry around [001] and the silicon crystal surface, oriented along (001). The contact center coincides with the

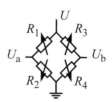

Fig. 2.2. Wheatstone bridge.

coordinate origin. To achieve a high rejection of thermal and intrinsic stress effects, the geometry of the sensor elements should fulfill the following requirements:

- All sensor elements should sense an equivalent thermal field, i.e. the thermal field possesses the same symmetry as the sensor element geometry.
- All sensor elements should be surrounded by equivalent neighboring structures.
- The design of the sensor elements is identical in terms of geometrical structure.

The above requirements imply the same geometry for all sensor elements in all directions. In reality, this requirement is difficult to fulfill, due to other design restrictions. By postulating a triaxial sensor design with outputs directly proportional to the forces in the directions of [110], [1$\bar{1}$0], and [001], the need of sensor element identity can be confined to sensor elements of the same axis. The postulated diagonal response renders post-recording diagonalization unnecessary. For low cross-coupling, temperature effects and other imperfections can be individually calibrated for each axis. This reduces the complexity of the measurement data handling, and thus increases the accuracy.

To achieve the uniaxial force sensitivity, the sensors make use of the symmetry of the Wheatstone bridge, of the contact and of piezoresistivity. Based on the linear approximation, the response of the sensor system is described by a set of three linear independent sensor signals

$$s_i(F_j) \equiv \frac{S_i}{U} = g_{ij}F_j,$$ (2.14)

that are normalized by the applied sensor voltage. The forces that are applied to the electro-mechanical contact zone C are defined by the integrals

$$F_j = \int_C \sigma_{jz} df.$$ (2.15)

The second order tensor g_{ij} is defined by equation (2.14) and subsequently referred to as *transductance tensor*. It relates the normalized sensor signal to the applied force vector F_j. For a full description of the sensor performance, the tilting moments of the contact should also be included in equation (2.14) [14]. These moments are coupled to the displacement of the contact, as described in [15]. Furthermore, the in-plane tilting moments M_x and M_y break the same symmetries as the tangential tractions F_x and F_y. Therefore, a sensor that is able to distinguish between tilting and tangential traction has also to elaborate the spatial distribution of the resulting stress fields.

Subsequent discussion demonstrates the possibility of designing a sensor with a diagonal transductance tensor under the condition of a point-symmetrical contact zone. The feasibility of such a sensor design is closely related to the stress field properties which result from an externally applied force. Consequently, the symmetry properties of the stress field have to be examined. An external force that is applied to the electro-mechanical contact, results in a stress field distribution

$$\sigma_{ij}[F](x). \tag{2.16}$$

The functional σ_{ij} is the solution of the equation of equilibrium for the specific boundary conditions at the surface of the half space for a given force vector F. As we only consider linear elasticity, the stress field of an arbitrary force is the linear combinations of the stress field induced by the components of the external force along the axis directions [110], [1$\bar{1}$0], and [001]. In other words,

$$\sigma_{ij}[F](x) = \sum_k \sigma_{ij}[F_k](x). \tag{2.17}$$

The stress field caused by one of these force components has to follow the symmetries of the applied force. This is often referred to as Curie's principle or symmetry principle [16]. This can be illustrated by the transformation rules for σ_{ij} for an applied force F_x under the various (001) plane-symmetry operations of the silicon crystal.

$m_{(110)}$: $$\sigma_{ij}[F_x](x, y) = -(-1)^{\delta_{ix}}(-1)^{\delta_{jy}}\sigma_{ij}[F_x](-x, y) \tag{2.18}$$

$m_{(1\bar{1}0)}$: $$\sigma_{ij}[F_x](x, y) = (-1)^{\delta_{ix}}(-1)^{\delta_{jy}}\sigma_{ij}[F_x](x, -y) \tag{2.19}$$

$m_{(100)}$, $m_{(010)}$, $4_{[001]}$: broken symmetry. $\hspace{3cm}$ (2.20)

The first minus sign of equation (2.18) is the result of the symmetry property of the force vector $F = (F_x, 0, 0)^T$ under reflection at the y-z-plane. The two other signs are the transformation property of the stress tensor. The symmetries of the stress functional for F_y is correspondingly derived. The force F_z does not break plane symmetries that are going through the center of the contact. The stress functional thus transforms like a covariant second-order tensor under all these plane symmetry operations.

The resistance change of a resistor results from the solution of the electric potential function $\Phi[\sigma(\bullet), \Delta T(\bullet)](x)$ on the sensor element area Ω_i. The dots indicate the place for the coordinate argument of the inner functionals. The electrostatic potential Φ has to satisfy the differential equation

$$\partial_i(\Lambda_{ij}\partial_j\Phi) = 0 \tag{2.21}$$

on the sensor element area for a given conductivity tensor Λ_{ij}

$$\Lambda_{ij} = (1 - \pi_{ijkl}\sigma_{kl} - \alpha_1\Delta T)/\bar{\rho}. \tag{2.22}$$

Both the piezoresistive coefficients π_{ijkl} and the stress field components σ_{ij} are given in the rotated coordinate system as defined in Fig. 2.1. Uniform Dirichlet boundary conditions are defined on a portion of the boundary $\partial\Omega$, called $\partial\Omega^D$. On the remaining portion of the boundary, $\partial\Omega^N$, vanishing Neumann conditions hold. These two portion cover the whole boundary:

$$\partial\Omega_i = \partial\Omega_i^D \cup \partial\Omega_i^N. \tag{2.23}$$

In a first step we consider ΔT having a rotational symmetry around the center of the contact. To distinguish between the resistor elements for the different axes, an additional suffix is added to the geometry: $\Omega_{i,j}$ is the geometry of the resistor element with index i along axis j.

The piezoresistive coefficients π_{ijkl} and the temperature coefficient of resistivity (TCR) α_1 are constants. The piezoresistive tensor π is obtained from Π by a transformation into the [110], [1$\bar{1}$0], and [001] primed coordinate system. The differential equation (2.21) is also employed for electrical device modelling. The parametric 2D device geometry, a 2D spatially varying stress field and the piezoresistive properties are the input parameters of the simulation.

We now define a functional that maps the electrostatic solution to the normalized change in resistance of the resistor k.

$$\mathfrak{R}_{k,i} \equiv \frac{\Delta R_{k,i}}{R} = f[\Omega_{k,i}, \sigma_{lm}[F](\bullet)](x) \tag{2.24}$$

The requirements at the beginning of this section demand the same nominal resistance R for the sensor elements along each direction. The normalized sensor signal s_i is expressed as a nonlinear function (Wheatstone bridge) of the resistor functional $\mathfrak{R}_{k,i}$

$$s_i = W(\mathfrak{R}_{k,i}) \equiv \left(\frac{1 + \mathfrak{R}_{2,i}}{2 + \mathfrak{R}_{1,i} + \mathfrak{R}_{2,i}} - \frac{1 + \mathfrak{R}_{4,i}}{2 + \mathfrak{R}_{4,i} + \mathfrak{R}_{3,i}} \right). \tag{2.25}$$

For small relative resistance changes the Wheatstone bridge equation can be linearized to

$$\tilde{s}_i \equiv \tilde{W}(\mathfrak{R}_{k,i}) = \frac{1}{4} \sum_k \alpha_k \mathfrak{R}_{k,i} \quad \text{with} \quad \alpha_k = \{-1, 1, 1, -1\}. \tag{2.26}$$

Linearized responses are henceforth marked with a tilde. Initially, the x- and y-force sensors are considered. The arrangement of the sensor elements is defined in Fig. 2.3. The sensor element arrangement of the x- and y-force sensor is fully compatible with the requirements listed at the beginning of the section. It is important to mention that the geometry of the sensor elements transform to each other by applying the symmetry operations, i.e.

$$\Omega_{\Gamma(k,\,i)} \equiv \Omega_{k,\,i}(\Gamma x) = \Omega_{l,\,i}. \tag{2.27}$$

The transformation Γ is a coordinate mapping corresponding to the specific symmetry operations that are displayed in Fig. 2.3 by solid lines. The transformation property of the geometry ensures that $\mathfrak{R}_{k,\,i}$ is independent of the tensor transformation properties of the stress components. Thus, only the symmetry property of the external force will change the sign of $\mathfrak{R}_{k,\,i}$ for the considered mirror operations (see Fig. 2.3). The resistor functional thus transforms according to

$$\Gamma \mathfrak{R}_{k,\,i} = \text{sgn}(\Gamma F)\mathfrak{R}_{k,\,i}, \tag{2.28}$$

where Γ is a plane symmetry transformation of the crystal symmetry group that is compatible with the symmetry properties of the applied force and the symmetry of the sensing element geometry.

As the x-force sensor and y-force sensor are equivalent except for a rotation of $\pi/2$, the following considerations are elaborated for the x-force sensor without loss of generality.

$$s_x = \frac{1 + P\mathfrak{R}_{1,\,x}}{2 + \mathfrak{R}_{1,\,x} + P\mathfrak{R}_{1,\,x}} - \frac{1 + P\mathfrak{R}_{3,\,x}}{2 + P\mathfrak{R}_{3,\,x} + \mathfrak{R}_{3,\,x}} \tag{2.29}$$

whereas P is defined by

$$P = \text{sgn}(\Gamma_{m_{(110)}} F). \tag{2.30}$$

If $P = -1$, equation (2.29) yields

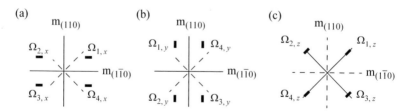

Fig. 2.3. Proposed sensor element arrangement for the x-force sensor (**a**), the y-force sensor (**b**), and the z-force sensor (**c**). Solid lines mark symmetry elements of the crystal that are also valid for the geometry of the sensor elements. Dashed lines are by the sensor geometry broken symmetry elements.

$$s_x = \frac{1}{2}(\Re_{3,x} - \Re_{1,x})$$

(2.31)

otherwise

$$s_x = 0.$$

(2.32)

Substituting the symmetry properties of equations (2.18) yields $g_{xj} = 0$ for $j \neq x$. Analogous symmetry properties can be applied to show that $g_{yj} = 0$ for $j \neq y$.

The sensor element arrangement as shown in Fig. 2.3c breaks the symmetry in $m_{(110)}$ and $m_{(1\bar{1}0)}$. Instead, the symmetry planes $m_{(100)}$ and $m_{(010)}$ are valid. Inserting the symmetry properties of the stress functional for a force F_z and the equivalence of the geometries in equation (2.25) or (2.26) yields $g_{zj} = 0$ for $j \neq z$. Thus, the sensor arrangement shown in Fig. 2.3 yields the desired diagonal transductance matrix g_{ij}, indicating a complete separation of forces in different axis directions for centered contacts.

The above considerations are based on the assumption of a perfectly centered contact, i.e. with sensor bridge and contact sharing the same center of symmetry. If the symmetry conditions are broken, cross-coupling between the force signals will result. An important case of symmetry violation are misplaced balls. The stress field of the contact has a nonlinear spatial dependence, dropping with increasing distance from the center of the contact. This nonlinear behavior can not be fully compensated by the sensor design and the offset between contact center and sensor center alters the sensitivity. An example of a sensitivity contour plot is shown in Fig. 2.4. The contour lines mark the positions of the sensor surface with equal sensitivity. An optimized sensor design should allow a large offset in position with minimal change in sensitivity. The quantities $r_{x\,2\%}$ and $r_{y\,2\%}$, defined as the minimal radius in x- and y-direction which leads to a sensitivity change of 2 %, have been used to compare the different sensor designs.

As the placement error susceptibility is an important design property, it is essential to examine the behavior of the transductance tensor for small placement errors. The following considerations are restricted to the linearized Wheatstone bridge function.

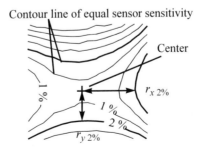

Fig. 2.4. Minimal placement error that can be tolerated for an accuracy of 2 %

In order to express the sensitivity as a function of the placement error $r = (x, y)^{\mathrm{T}}$, the transductance tensor

$$g_{ij}(r) = \overset{0}{g_{ij}} + x_k g_{ijk} + \frac{1}{2} x_k x_l g_{ijkl} + O(r^3) \qquad (2.33)$$

is expanded around the center position in ascending powers of x_i. Variations of the contact position along the z-axis (as might result from a change of the pad metallization thickness) are not considered, as the material layer system would have to be included for a correct description of a variation in z.

The elements g_{ijk} and g_{ijkl} of equation (2.33) are the first and second order derivatives of g_{ij} defined by

$$g_{ijk} \equiv \partial_k g_{ij}\big|_{r=0} \quad \text{and} \qquad g_{ijkl} \equiv \partial_k \partial_l g_{ij}\big|_{r=0}. \qquad (2.34)$$

The expansion is justifiable as long as the sensor elements are distant from the contact zone. In order to be able to determine the symmetry-induced properties of g_{ijk} and g_{ijkl} we need to know how g_{ij} behaves when stepping out of the center by an infinitesimal distance. This can be achieved by examining the properties of the transductance tensor g_{ij}, whether it is an odd or even function of x and y respectively.

The commutative diagram of Fig. 2.5 schematically depicts the use of the transformation properties of the force vector and the Wheatstone bridge to extract the symmetry properties of g_{ij}. The contact is off-centered by a distance Δy. A force F_y is applied on the contact. The direction of the force is indicated with an arrow. The physics of this system does not alter if the entire system is mirrored with respect to the x-z-plane. By changing the sign of the applied force, the force vector is pointing again in the same direction as for the initial configuration (Fig. 2.5a). A remapping

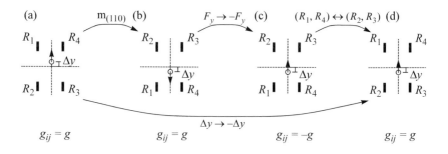

Fig. 2.5. If g_{ij} is an even function of y, the sign of g_{ij} is equal for a positive and negative value of Δy. The sensor configuration for a negative value Δy is equivalent to a sign change of the force and a remapping of the Wheatstone bridge.

of the resistor elements of the Wheatstone bridge yields a configuration (Fig. 2.5d) that has a large similarity to configuration (a) differing only in the negative displacement Δy of the contact center. Thus, the sensor signal change due to a displacement along the y-axis is equivalent to a force direction change and a remapping of the Wheatstone bridge elements.

We first consider the symmetry of the Wheatstone bridge functional $\tilde{W}_i$. A summary of the polarity under the different symmetry operations is found in Table 2.3.

Table 2.3. Symmetry properties of the linearized Wheatstone bridge under mirror operations. A dash marks the absence of symmetries.

Symmetry of the linearized Wheatstone bridge	$m_{(110)}$	$m_{(1\bar{1}0)}$	$m_{(100)}$	$m_{(010)}$
$\tilde{W}_x$	odd	even	—	—
$\tilde{W}_y$	even	odd	—	—
$\tilde{W}_z$	—	—	even	even

Furthermore, the symmetry properties of the value of each resistor have to be determined. From the requirements that the resistor geometries transform to each other under mirror operations, the resistivity is independent of the tensor transformation properties of the stress field components. Therefore, the resistance change only depends on the symmetry or antisymmetry property of the applied force, as summarized in Table 2.4.

Table 2.4. Symmetry properties of the resistance change caused by a contact with applied force F.

Symmetry of the resistor functional	$m_{(110)}$	$m_{(1\bar{1}0)}$	$m_{(100)}$	$m_{(010)}$
F_x	odd	even	—	—
F_y	even	odd	—	—
F_z	even	even	even	even
$F_x + F_y$	—	—	—	odd
$F_x - F_y$	—	—	odd	—

The combination of the symmetry of the resistance change and the Wheatstone bridge yields the symmetry properties of g_{ij}, as listed in Table 2.5. It is remarked that the composition of two odd functions is even, whereas the composition of odd and even yields an odd function.

Table 2.5. Even/odd symmetry of the combination of the by the force F caused sensor element response and the Wheatstone bridge.

Symmetry of g_{ij}	Direction of misalignment	F_x	F_y	F_z
$\tilde{s}_x$	[110]	even	odd	odd
	[1$\bar{1}$0]	even	odd	even
$\tilde{s}_y$	[110]	odd	even	even
	[1$\bar{1}$0]	odd	even	odd
$\tilde{s}_z$	[110]	—	—	even
	[1$\bar{1}$0]	—	—	even

Using the symmetry properties of Table 2.5 in the directions x and y, the lowest-order non-zero terms of the transductance expansion can be determined. Furthermore, the sensitivity surface is either even or odd in x or/and y, respectively.

The symmetry of a perfectly centered contact implies that only diagonal terms involve a constant value in the expansion of the transductance. Thus, an off-diagonal term that is even both in x and y must be at least quadratic, of the form $g_{ijxx}x^2 + g_{ijyy}y^2$.

The cross-talk of F_x in s_y is odd both in x and y. Thus, the lowest order term in the expansion of the transductance is proportional to xy with its coefficient g_{yxxy}. From the geometrical equivalence follows that $g_{xxxx} = g_{yyyy}$, $g_{xxyy} = g_{yyxx}$, and $g_{yxxy} = g_{xyxy}$. A form like $ax + by$ can be ruled out as either the first or second summand would be even in the other coordinate. The cross-talk of F_z in s_x is odd in x and even in y and therefore proportional to x. The cross-talk of F_x and F_z in s_z is extracted from the behavior of $F_x + F_y$ in Table 2.4. In combination with the property of the Wheatstone bridge follows that the lowest-order term of the cross-talk is proportional to $x + y$. Table 2.6 summarizes the lowest-order terms of the sensitivity and cross-talk surfaces.

Table 2.7 exemplarily shows contour plots that correspond to the functional behavior tabulated in Table 2.6. The parabolic sensitivity surfaces of the diagonal elements are based on equal sign of g_{iixx} and g_{iiyy}. An opposite sign results in a saddle surface. Experimental verifications of these sensitivity surfaces are found in Sect. 4.4.

In conclusion:
- No cross-coupling exists for an exactly centered circular contact.
- Characteristics of the x- and y-force sensors are identical for equal sensor element geometry and positioning.
- The iso-sensitivity surface of s_i for a force F_i with $i = \{x, y\}$ is purely parabolic.
- Cross-coupling terms are dominated by terms that are bilinear with the contact misplacement.

Table 2.6. Lowest-order terms of sensitivity/cross-talk surfaces.

	F_x	F_y	F_z
$\tilde{s}_x$	$g_{xx} + g_{xxxx}x^2 + g_{xxyy}y^2$	$g_{xyxy}xy$	$g_{xzx}x$
$\tilde{s}_y$	$g_{yxxy}xy$	$g_{yy} + g_{yyxx}x^2 + g_{yyyy}y^2$	$g_{yzy}y$
$\tilde{s}_z$	$g_{zxx}x + g_{zxy}y$	$g_{zyx}x + g_{zyy}y$	g_{zz}

Table 2.7. Schematic contour plots of functions defined in Table 2.6.

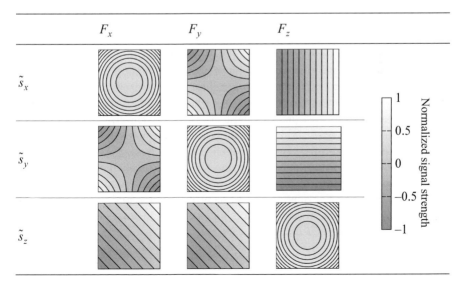

2.1.3   Stress Fields

Extensive literature exists about encapsulation and thermal expansion induced stress fields. Various packaging test chips for the experimental identification of failure sources during the packaging process and accelerated failure tests have been developed. Packaging test chips optimized for measuring surface stress fields, are generally based on rosette shaped piezoresistors oriented with respect to particular axes to measure the full set of stress components [3].

On the opposite there is few information available about stress fields that origin from electro-mechanical contacts of the packaging process [17, 18]. The stress fields of electro-mechanical contacts are strongly localized at the position of the contact. The field of mechanical contact theory offers analytical solutions for specific contact problems. Parameter dependence can be easily studied with these analytical solutions. In contrast, finite element method (FEM) models offer the possibility to account for the spatial complexity of real sensor structures.

The design of the sensors incorporates a trade-off between signal strength and susceptibility to changes in the boundary condition of the contact. For instance, a deviation in the contact size or contact position relative to the sensor will diminish the sensor accuracy. Furthermore, different mechanical contact conditions significantly alter the stress field close to the area of contact [19]. In order to optimize the sensor performance, the knowledge of the stress fields of these contacts is essential. Both, an analytical and a FEM approach will be used to gain insight into the stress field distribution.

Contact Conditions. During the impact phase of the wire bonding process, the ball is plastically deformed due to a strong normal force. This results in a circular contact zone between ball and bonding pad. Simulations of this deformation process were performed by different authors [18, 20, 21]. The insight from these thermo-plastic deformation simulations is a viable explanation of the ring-shaped bond zones observed for thermo-compression bonds [22]. During the impact phase tangential traction as a result of the ultrasound oscillation of the bonding tool is absent or negligible compared to the applied normal force. The deformation and the substrate temperature are the dominant causes of bond growth during the impact phase strongly affected by the surface quality and cleanness. Mechanical cross coupling can result in an additional low-frequency force in y-direction. The exploration of the impact deformation and related bonding effects is not scope of this work.

The contact state varies during the bond process. After impact deformation, the gold ball is pressed onto the pad by the applied normal force F_z. During the ultrasound bonding time, an additional tangential force F_T is applied that can result in gross sliding [23] between the ball and the pad under the condition of $F_T > F_{frict}$, with the friction force F_{frict} defined by the Coulomb friction law $F_{frict} = \mu F_z$.

With ongoing bonding time, friction is inhibited due to bond growth. Thus, the bonded contact experiences other boundary conditions than the sliding or sticking contact. Due to the high forces during the bonding process, plastic deformation is present, resulting in a complex time-history dependent problem. An extended overview on elastic-plastic contact problems is found in [19].

Wire bonding and flip-chip contact materials possess a relatively low yield strength σ_{yield} of about 170 MPa and 30 MPa, respectively. Therefore, high stress fields at the contact boundaries will immediately result in plastic flow, spreading the stress field across the whole contact zone. These varying boundary conditions will change the stress fields near the contact zone.

In other applications (see Sect. 3.2), a ceramic capillary is in direct contact with the passivation of the chip. Both, the ceramic material and the Si_3N_4 of the top layer passivation are materials with a large Young's modulus of about 350 GPa and 250 GPa, respectively. As a consequence, stress fields at the contact border are not bound by plastic effects. These contacts are close to the problem of a rigid cylinder contact on an elastic half space with its divergent stress behavior at the contact boundary [24].

At distant positions, changes in the actual boundary conditions of the contact result in a marginal effect on the stress, strain and displacement field if the resulting force and moments remain constant. This behavior is also known as Saint-Venant's principle [19]. Therefore, by placing the sensor elements distant from the contact will lower the susceptibility to changes in the contact geometry and the contact state. This is especially important for the wire bonding application.

To get a qualitative understanding of the stress field distribution, the boundary condition at the contact is approximated by a rigid cylinder contact problem for the subsequent sensor design considerations. As the interest lies in the local stress fields around the contact and not in displacements, the chip die can be approximated by a half space as long as the sensor is not close to the chip border.

An arbitrary force distribution on the surface of an elastic half space can be expressed by a superposition of the solution of a point load $P = F\delta(r)$ by using Green Integrals. The Green tensor is derived from the displacement vector u as e.g. given for isotropic materials in [25]. A straightforward integration of the Green tensor is difficult for most of the boundary conditions. The transformation of the problem to potential solutions offers an alternative approach.

Stress Field of a Contact Under Normal Force. For a rigid cylinder pressed frictionless with a normal force F_z against an isotropic half space, the boundary conditions are given by a circular contact area of radius r_c displaced by u_z. Stress fields outside and the shear stress fields inside the contact zone vanish at the surface.

$$\begin{array}{ll} u_z\big|_{z=0} = u_0(F_z) & r \leq r_c \\ \sigma_{zz}\big|_{z=0} = 0 & r > r_c \end{array} \qquad \begin{array}{l} \sigma_{xz}\big|_{z=0} = 0 \\ \sigma_{yz}\big|_{z=0} = 0 \end{array} \qquad (2.35)$$

The vanishing shear stress fields correspond to a friction-less contact. On the basis of the Papkovitch-Neuber equation [26]

$$2\xi u = -(4-v)\Psi + \nabla(v \cdot \Psi + \phi) \tag{2.36}$$

that transforms the differential equation of equilibrium of the displacement vector u into an equation that contains two harmonic functions Ψ and ϕ, and the stress strain relation

$$\sigma_{ij} = \lambda\varepsilon_{kk}\delta_{ij} + 2\xi\varepsilon_{ij}, \tag{2.37}$$

the calculation of the stress fields results in finding harmonic functions that meet the specific boundary conditions (2.35). Equation (2.37) is Hook's law for isotropic materials with the Lamé's constants λ and ξ related to the Young's modulus E and Poisson's ratio v by

$$\xi \equiv \frac{E}{2(1+v)} \text{ and } \lambda \equiv \frac{Ev}{(1+v)(1-2v)} = \frac{2\xi v}{(1-2v)}. \tag{2.38}$$

As the imaginary part $\Im$ of a harmonic function is itself harmonic, it is shown in [26] that the function φ, defined by

$$\varphi = k \cdot \Im\{(z + ir_c)\log(R_a + z + ir_c) - R_a\} \text{ with } R_a \equiv \sqrt{r^2 + (z + ir_c)^2}, \tag{2.39}$$

$$\phi = (1-2v)\varphi \text{ and } \Psi = e_z\frac{\partial\varphi}{\partial z} \tag{2.40}$$

fulfills the boundary conditions in (2.35). This harmonic function φ, derived from singular spherical harmonics by the transformation $z \to z + ir_c$, can now be used in combination with (2.37) to calculate the stress field in the half space. The constant

$$k = \frac{2\xi}{\pi(v-1)}u_0 \tag{2.41}$$

is found by integration of the stress field σ_{zz} over the contact zone.

Figures 2.6 and 2.7 show the isosurfaces of the stress field σ_{zz} and σ_{xy} calculated with equations (2.36) to (2.41) for a circular contact with radius $r_c = 25$ μm and an applied normal force of 1 N. The parameters describing the mechanical properties of silicon are found in Table 2.8. The large stress fields at the border of the contact zone are due to the assumption of a rigid contact resulting in divergent stress fields at the edge. The σ_{zz} stress field is mainly restricted to the silicon underneath the contact zone. A normal force applied to the contact zone also results in stress field components that spread over the silicon surface outside the contact zone as e.g. seen in Fig. 2.7 for σ_{xy}.

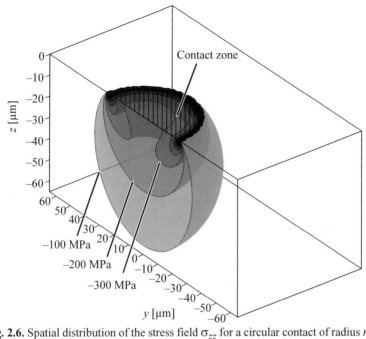

Fig. 2.6. Spatial distribution of the stress field σ_{zz} for a circular contact of radius $r_c = 25$ μm under a normal force of 1 N. Shown are isosurfaces with stress values -100 MPa, -200 MPa, -300 MPa, -400 MPa, -500 MPa, -600 MPa, and -700 MPa.

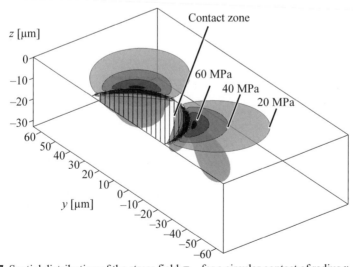

Fig. 2.7. Spatial distribution of the stress field σ_{xy} for a circular contact of radius $r_c = 25$ μm under a normal force of 1 N. Shown are isosurfaces with stress values ±20 MPa, ±40 MPa, ±60 MPa, and ±80 MPa.

Stress Field of a Contact under Tangential Traction. The exact solution for transversal traction of a rigid cylinder bonded to a transversal isotropic half space is found in [15]. The boundary conditions of this bonded contact are given by

$$
\begin{aligned}
u_x(x, y, 0) &= u_x & r \leq r_c & \qquad \sigma_{xz}\big|_{z=0} = 0 & r > r_c \\
u_y(x, y, 0) &= 0 & r \leq r_c & \qquad \sigma_{yz}\big|_{z=0} = 0 & r > r_c \\
u_z(x, y, 0) &= -\gamma x & r \leq r_c & \qquad \sigma_{zz}\big|_{z=0} = 0 & r > r_c.
\end{aligned}
\tag{2.42}
$$

In contrast to the normal force contact, the boundary conditions inside the contact area are no more specified by vanishing shear tractions, but by a tangential displacement u_x and a tilt around the y-axis described by γ.

An approximative solution of the stress field can be derived under the assumption of a parabolic shear stress field at the contact surface, as mentioned in [24]. Based on this approximative boundary condition and the potential function as used in [27], an explicit form of the stress field is obtained.

FEM simulations were performed with ANSYS$^\circledR$ to account for the finite dimension of the die and the layer stack of silicon dioxide / silicon nitride on top of the silicon. For stress field calculation, static boundary conditions are applied on the contact at the center of a silicon die with dimensions of 3000 μm side length and 620 μm depth. Table 2.9 summarizes the mechanical properties of the materials, transformed to the [110], [1$\bar{1}$0], and [001] coordinate system. Simulations that approximate silicon with isotropic behavior are based on the Young's modulus $E = 169.2$ GPa and the Poison ratio $\nu = 0.026$. A detail view of the used mesh in the vicinity of the contact zone is shown in Fig. 2.8. The mesh is based on tetrahedral and brick shaped Solid95 elements. The transition zone is formed by pyramid shaped elements. The combination of two different element types was chosen to achieve a dense uniform mapped meshing in the read-out plane of the stress field with a moderate total element count.

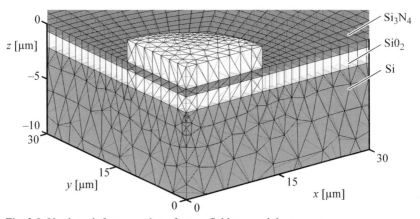

Fig. 2.8. Used mesh for extraction of stress fields around the contact.

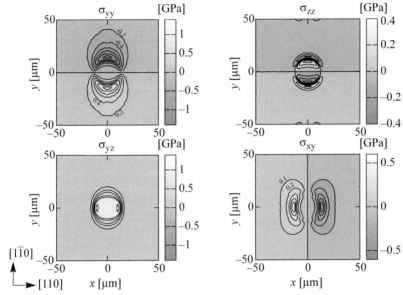

Fig. 2.9. Stress fields 2.2 μm below the surface for a rigid bonded cylinder with $r_c = 12.5$ μm contact radius under tangential traction of 1 N in y-direction. The tilt angle is fixed to zero which results in a tilting moment. The resulting tangential displacement of the contact zone is 274 nm.

Table 2.8. Mechanical properties used for stress field calculation. Coordinate system orientation as defined in Fig. 2.1.

	Young's modulus [GPa]	Shear modulus [GPa]	Poisson's ratio	Ref.
Si	$E_x = 169.2$	$G_{yz} = 79.6$	$\nu_{xy} = 0.062$	[28]
	$E_y = 169.2$	$G_{xz} = 79.6$	$\nu_{xz} = 0.362$	
	$E_z = 130.2$	$G_{xy} = 50.9$	$\nu_{yz} = 0.362$	
Si$_{isotropic}$	169.2		0.062	
SiO$_2$ (thermal)	65		0.17	[29]
SiO$_2$ (PECVD)	65		0.17	[29]
Si$_3$N$_4$ (PECVD)	250		0.254	[29, 30]

The boundary conditions of the (110) and ($1\bar{1}0$) planes are selected according to the symmetry of the applied forces. For stress field extraction in the sensor plane, appropriate mesh size becomes important, as spatially fluctuations in the stress field directly affect the accuracy of the parametric sensor optimization. Fig. 2.9 exemplarily displays an extracted stress field in the silicon caused by a rigid bonded contact under tangential traction. The CMOS process layers are taken into account by a SiO_2 layer (consisting of field oxide, contact oxide, and via oxide) that is covered by Si_3N_4. The stress field is extracted from the elements of the silicon at the junction between the silicon and the silicon oxide layer. Only the stress field components with substantial strength are plotted.

High stress field gradients are visible at the contours of the contact. These areas are moreover the zone of large stress field values. Sensors that graze these areas can be optimized for sensitivity but are also prone to changes in the contact position, contact geometry, and contact type. This trade-off between sensitivity and accuracy is extensively discussed in Sect. 2.2.

The stress distribution for arbitrary z-positions is shown in the Figs. 2.10 and 2.11 for different isosurfaces of half of the silicon chip. The cross-section of the

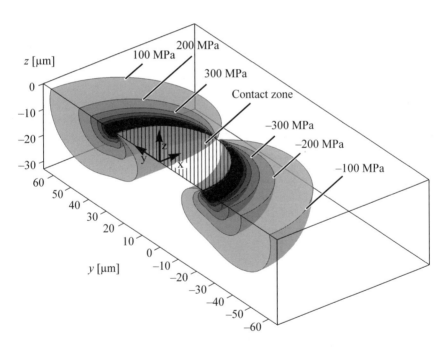

Fig. 2.10. Spatial distribution of the stress field σ_{yy} for a circular contact of radius $r_c = 25$ μm under tangential traction of 1 N in y-direction. Isosurfaces of the stress values ±100 MPa, ±200 MPa, ±300 MPa, ±400 MPa, ±500 MPa, and ±600 MPa. The cross-section of the zy-plane is shown.

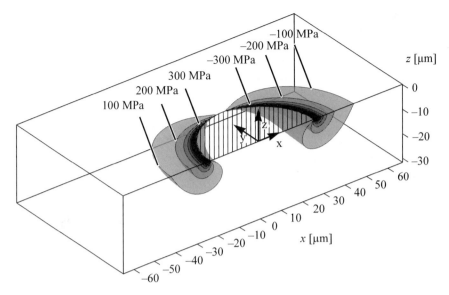

Fig. 2.11. Spatial distribution of the stress field σ_{xy} for a circular contact of radius $r_c = 25$ μm under tangential traction of 1 N in y-direction. Potential surfaces of the stress values ±100 MPa, ±200 MPa, ±300 MPa, ±400 MPa, ±500 MPa, and ±600 MPa. The cross-section of the xz-plane is shown.

yz-plane is shown in Fig. 2.10. The shown stress field is based on a analytical calculation for an isotropic body under absence of any layers of different materials. The circular contact zone is marked with a hatched area.

As the σ_{xy} stress field in the yz-center-plane is zero as consequence of the symmetry properties (see Sect. 2.1.2), the chip is cut in the xz-plane in Fig. 2.11. A comparison between σ_{yy} and σ_{xy} indicates larger stress values for σ_{yy} for the same applied tangential force and distance from center.

2.1.4   Intrinsic Stress Fields

The processing of integrated circuitry makes use of successive deposition of layers that are structured. Differences in the thermomechanical properties of the layer materials and deposition techniques result in mechanical stress in the layers varying with temperature changes [31]. Furthermore, nonlinear behavior of the material, e.g. the onset of plastic deformation, will result in stress relaxation or hysteresis [32]. Effects that correlate with the geometry of the layered structure can be inhibited by making use of the symmetry for the sensor design.

2.2   Ball Bond Sensor

Packaging processes generally include process steps stressing the integrated circuitry beyond its normal operation conditions, involving e.g. high process temperatures, electrical noise, and high mechanical stress. Normal circuitry is not in an operational condition during these processes. In contrast, microsensors that are used for in situ and real-time characterization of the packaging processes have to operate under these harsh conditions, which demands robustness of the sensing system. The subsequent paragraph lists the requirements for a sensor system used for in situ packaging process inspection.

Substrate temperature has been known to have a large influence on the bond strength [33] of wire bonds. It is known that the bond strength of Al-Au contacts mostly improves with increasing substrate temperature. However, the applicable maximum temperature can be restricted by the substrate type [34]. As an example, the commonly used plastic ball grid array (PBGA) substrates based on bismaleimide triazine (BT) have a glass transition temperature T_g between 180 °C and 190 °C. For specific applications, bonding may also be performed at ambient temperature. Covering a large partition of the wire bonding temperature range from ambient temperature up to 250 °C demands for a high temperature stability of the sensor system. The use of circuitry substrates allowing higher operation temperatures, such as e.g. silicon on insulator (SOI) could increase the temperature range, but also restricts the sensor system to these substrate types with higher sensor production costs.

In addition to the high temperature stability, the sensor should be able to withstand large stress fields at the contact. A normal force of $F_z = 500$ mN applied with a ceramic type imprint capillary results in a σ_{zz} stress field that exceeds 1 GPa at its maximum. In plane stress fields around such contacts reach a similar strength (see Fig. 2.15). For higher loads, surface damage is observed that drastically diminishes the sensor's lifetime. For lower loads wear traces are visible on the surface. Drift of sensor offset is observed for high contact forces if the sensing elements are integrated directly under the contact zone.

As the sensor is used in an industrial environment and measurements are performed in situ during the packaging process, the sensor should be insensitive to electrical noise. This includes especially the insensitivity against capacitive coupling between the test pad and the sensor elements.

In order to characterize a real bonding process, the mechanical and structural behavior of the pad should not be altered by the implementation of the sensor. To resolve the individual oscillations of the harmonics of the ultrasound (typical frequency 130 kHz), the frequency range of the sensor should exceed 1 MHz. For packaging characterization, high repeatability of measurements is prerequisite in order to collect statistically relevant data about process parameter dependence and process stability.

The sensor design approaches can be grouped into three classes according to their geometrical arrangement in relation to the test pad. Figure 2.12 shows a schematic top view of a layout with sensor elements adjacent to the test pad, structured sensor elements that overlap the test pad, and a plain sensor structure overlapping

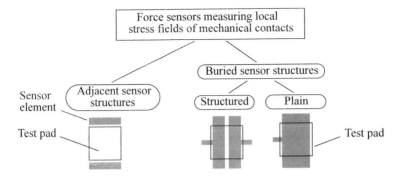

Fig. 2.12. Schematic top view of the different sensor design approaches.

the test pad. Buried sensors were first reported in [35]. A comparison between the different sensor structures can be found in Table 2.9. Integration of the sensor elements in regions of strong stress fields (e.g underneath the contact zone) results in a high sensor response but it also entails large stress field gradients which diminish the repeatability. In order to achieve high measurement repeatability and gain freedom in the test pad design, adjacent sensors have been designed [36]. This novel design approach of measuring local stress fields enables the coexistent integration of sensors that measure uniaxial forces in all three axes. Therefore, theses sensors are also named *xyz-force sensors*. The next chapter will summarize the design considerations for this class of sensors.

Table 2.9. Comparison of different design approaches as shown in Fig. 2.12. The optimal values are given in brackets.

	Structured buried sensor	Plain buried sensor	Adjacent sensor
Placement sensitivity (low)	high	medium	low
Pad design (unchanged)	altered	altered	unchanged
Sensitivity (high)	high	medium	medium
Simultaneous multi-axes measurement (yes)	no	no	yes
Wheatstone bridge type (full)	full	half	full

2.2.1   XYZ-Force Sensor

Design. Based on the symmetry considerations of Sect. 2.1, the sensor elements of the Wheatstone bridge (see Fig. 2.2) can be grouped in different arrangements. Figure 2.13 exemplarily lists the main sensor design families. The sensitivity of a sensor arrangement is given by the stress field strength at the sensor position and the piezoresistive properties of silicon.

We first consider the design of sensors for forces tangential to the chip surface. The stress field calculations in Sect. 2.1 (see Fig. 2.9) indicate beside σ_{yz} (which can not be accessed by in-plane sensors) significant stress field components σ_{yy} and σ_{xy} outside the contact area. Figure 2.14 plots the strength of these stress fields as function of the distance from the center for a tangential traction of 1 N. The peak values of the stress components σ_{xx}, σ_{yy}, and σ_{xy} are observed close to the boundary of the contact zone. The strength of the stress field component σ_{yy} is significantly higher than that of the stress field σ_{xx} for a tangential traction in y-direction. From the piezoresistive theory in combination with typical values of the piezoresistive coefficients as reported in Table 2.2, it follows that line-shaped p^+ diffused resistors aligned in $[1\bar{1}0]$ direction are mainly sensitive to the difference between σ_{yy} and σ_{xx} due to the dominant influence of π_{44}. This characteristic is used for the p^+ xyz-force sensor as shown in Fig. 2.13a. As the stress field is antisymmetric with respect to the $(1\bar{1}0)$ plane, a full Wheatstone bridge configuration can be achieved by placing two sensing elements on each side. Figure 2.13a schematically shows the four numbered resistor elements R_1^y to R_4^y connected as defined in Fig. 2.2. The arrangement of the x-force sensing elements is the same as that of the y-force sensing elements except for a rotation by $\pi/2$. The bar in Fig. 2.14 marks the sensor element position of the final design. The position of the sensor element has to be found by bal-

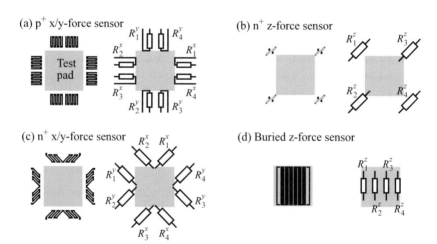

Fig. 2.13. Schematic of the sensors based on p^+ and n^+ diffused piezoresistors and their equivalent resistor network. The resistors are connected in a full Wheatstone bridge configuration as defined in Fig. 2.2.

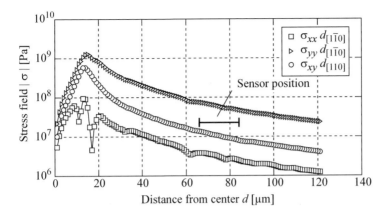

Fig. 2.14. Stress field in anisotropic Si covered with SiO_2 and Si_3N_4 layers calculated by FEM. A tangential force of 1 N is applied in $[1\bar{1}0]$ direction on a circular contact area with radius $r_c = 12.5$ μm. The stress fields σ_{xx} and σ_{yy} are plotted along axis $[1\bar{1}0]$, whereas σ_{xy} is plotted along axis $[110]$.

ancing of sensitivity versus placement error susceptibility and will be considered in the second part of this section.

While σ_{yy} dominates the stress field of a contact under tangential traction in y-direction, i.e. in the direction of the applied force, the shear stress σ_{xy} dominates in the x-direction. This stress field component σ_{xy} is smaller than the principal stress σ_{yy}. From equation (2.7) to (2.10) follows that highest sensitivity is achieved for sensor elements aligned in [100] or [010] direction.

Piezoresistors based on n$^+$ diffused elements have higher sensitivity to σ_{xy} stress fields. Consequently, the n$^+$ x/y-force sensor based on slanting serpentine resistor elements is selected. As the resistance change of [100] and [010] resistors has opposite sign, a full Wheatstone bridge can be formed as shown in Fig. 2.13c. It is to be mentioned that the y-force sensing elements are placed to the left and right of the test pad in contrast to the x/y-force sensor based on p$^+$ diffused resistors.

The stress field of a contact under a normal force fundamentally differs from the stress field of a tangential force. Therefore, the sensor elements of a z-force sensor are arranged in a different way. The next paragraphs summarize the design considerations for the z-force sensor. Figure 2.15 shows the stress field of a circular contact under a normal force of 1 N. A large σ_{zz} stress field predominates directly under the contact for an applied normal force, but vanishes rapidly outside the contact zone. Buried sensor structures sensitive to σ_{zz} stress can be used to sense normal forces. For this type of sensor, reference resistors are needed to form a full Wheatstone bridge. Resistors R_1 and R_4 in Fig. 2.13d are used as reference for the sensing resistors R_2 and R_3.

The stress field components σ_{xx}, σ_{yy}, and σ_{xy} are existent directly under the contact as well as outside the contact zone. From the symmetry properties of the contact

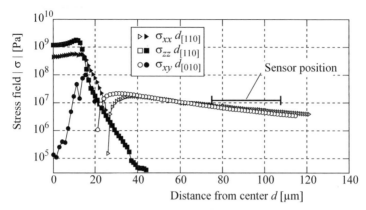

Fig. 2.15. Stress field in anisotropic Si covered with SiO_2 and Si_3N_4 layers calculated by FEM. A normal force of 1 N is applied on a circular contact area with radius $r_c = 12.5$ μm. Blank and filled markers stand for tensile and compressive stress values, respectively. The stress fields σ_{xx} and σ_{zz} are plotted along axis [110], whereas σ_{xy} is plotted along axis [010].

follows that except for changes due to the anisotropy of silicon the principal stress fields and the shear stress field show similar strengths outside the contact area. The stress field components σ_{xx}, σ_{yy}, and σ_{xy} are maximal in [110], [1$\bar{1}$0], and [010] direction, respectively. Either p^+ or n^+ piezoresistors can be selected to sense the normal force through principal or shear stress components, respectively. Figure 2.13b shows a z-force sensor based on n^+ diffused piezoresistors arranged along the [010] direction, i.e. a sensor sensitive to the σ_{xy} stress.

Electromechanical Simulation. Based on the calculated stress fields the electrical sensor response is determined with a 2D FEM solver by solving the electrostatic potential equation for a spatially varying stress field. The 2D FEM simulation is based on equations (2.21) and (2.22). The geometry of the considered sensing element structure is fully parameterized and allows a calculation of the sensor signal strength as function of selected parameters. The stress field input for the FEM model is a two dimensional grid of stress field values. The sensor element is described as a set of trapezoid shaped paths connected in parallel or in series. The simulation flow is schematically shown in Fig. 2.16. An appropriate accuracy of the piezoresistive coefficients and the stress fields is indispensable for reliable simulation result.

Optimization of Sensor Element Arrangement. Once the sensor element distribution is selected, mainly two degrees of freedom in the sensor geometry remain, the spacing w (distance between the two neighboring elements as shown in Fig. 2.20) and the distance of the sensor elements from the center d (see Fig. 2.18). Selecting these two parameters results in a trade-off between sensor sensitivity, geometrical dimensions, and placement error susceptibility. Figure 2.17 and

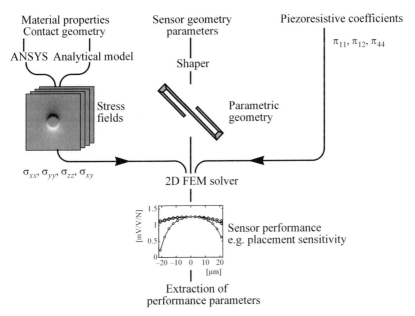

Fig. 2.16. Simulation flow for extraction of the sensor sensitivity for a given stress field, piezoresistive coefficients, and a sensor geometry.

Fig. 2.18 show the sensitivity g_{yy} and placement dependence parameters $r_{x\,2\%}$ and $r_{y\,2\%}$ as function of the distance d for the p$^+$ xyz-force sensor design. The calculations were performed with different degrees of model simplification. Figure 2.17 shows the decrease of the sensitivity g_{yy} for increasing distance d from the center. The FEM model with isotropic material parameters of silicon matches the analytical

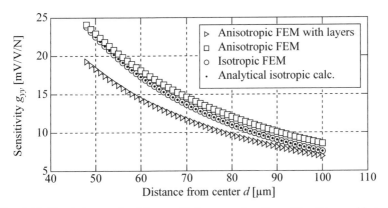

Fig. 2.17. Simulation result of the sensitivity g_{yy} as function of the distance d between sensor elements and center of the sensor. Radius of the contact $r_c = 12.5$ μm.

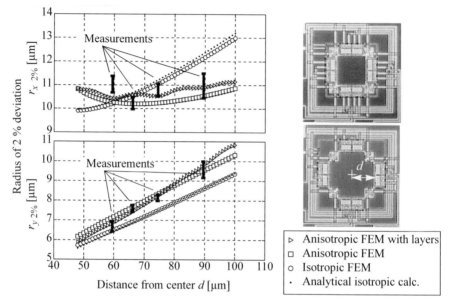

Fig. 2.18. Measurements and simulations of the radius of placement sensitivity of the y-force sensor due to a varying position of the sensor elements from the center. The micrographs on the right show the sensors with varied distance d. Contact radius is $r_c = 12.5$ μm.

isotropic calculation. The sensitivity for a silicon die covered with a SiO_2 and Si_3Ni_4 layer is significantly lower. The calculated sensitivity values are to be compared with the calibration of the sensors in Sect. 4.2.

The extracted parameters $r_{x\,2\%}$ and $r_{y\,2\%}$ are particularly well suited for comparing simulation and measurement results, due to their high sensitivity to model simplifications and independence of the sensor calibration. The micrographs on the right side of Fig. 2.18 show sensor designs with varied distance of the sensor elements to the center. Comparison of the values of the different simulation models and the measurement yields an incorrect trend for simulations based on isotropic approximations. Measurements and simulations based on anisotropic silicon show a constant $r_{x\,2\%}$ and $r_{y\,2\%}$ increases approximately linear with distance d. For an explanation of the measurement procedure to extract $r_{x\,2\%}$ and $r_{y\,2\%}$ refer to Sect. 4.2. Variation of the spacing w shows that the curvature of the sensitivity surface is changing the sign with increasing w. Figure 2.20 shows simulation results for a sensor design with $d = 59.5$ μm. The radii $r_{x\,2\%}$ and $r_{y\,2\%}$ are changing sign at a spacing $w = 30$ μm and $w = 65$ μm, respectively. The decrease in the placement error susceptibility has to be paid with a decrease in signal strength. This behavior can be experimentally verified. Figure 2.19b shows a sensitivity surface for a sensor with 50 μm element spacing. The sensitivity surface has changed from a saddle shaped surface (see Fig. 2.4) to a parabolic sensitivity surface. Similar design parameter optimization can be performed for the other sensor types.

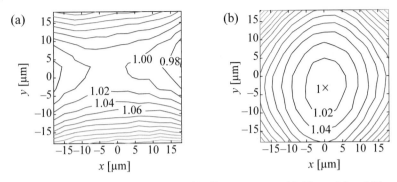

Fig. 2.19. Measurement of $s_y(x, y)/s_y(0, 0)$ of a sensor with 20 μm (**a**) and 50 μm (**b**) element spacing. The sensitivity surfaces are to be compared with the simulated data in Fig. 2.20.

In conclusion, increasing the distance d or the sensor element spacing w yields an improved insensitivity to placement variations. However, this has to be paid with a lower sensor sensitivity. Furthermore, cross-talk properties will degenerate. Therefore, a small spacing w and a moderate distance d is chosen for the xyz-force sensor.

Integration into the CMOS Process. On-chip integrated microsensors for packaging test chips can only be used once for bond process inspection (disposable sensor). Consequently, to achieve low production cost and facilitate process transfers of a packaging test chip, a standard process was selected for fabricating the sensors. The sensor design is based on a commercially available double metal, single poly-

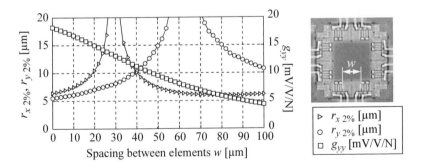

Fig. 2.20. Simulation of $r_{x\,2\%}$ and $r_{y\,2\%}$ of the y-force sensor as function of the spacing w between the sensor elements. The micrograph on the right shows a sensor with 50 μm spacing between the sensor elements. The sensitivity contour lines of this sensor are found in Fig. 2.19b.

silicon 0.8 μm CMOS process. The process is carried out on 6 inch lightly p-doped (100) wafers. In order to achieve best possible temperature insensitivity, highly doped diffusions are selected for piezoresistive sensing. During the packaging process rather high local stress fields are generated. As the sensor is used in a wide temperature range, improved temperature stability will overweight the lower sensitivity. Therefore, the source/drain implantations of the CMOS process are chosen for the sensing elements. Figure 2.21 shows a schematic cross-section of the pad and the sensor structure. The p^+ source/drain implantation is formed in an n-well. The n-well is connected with an n^+ contact to the positive supply voltage (VDD) of the chip. The resulting pn-junction forms the geometry of the piezoresistors. The junction depth of the p^+ diffusion is about 0.35 μm. Due to the high doping level of the p^+ diffused layer, the depletion layer thickness of the sensing layer is only slightly dependent on the bias voltage. Sensor structures based on n^+ source/drain implantations are integrated into the self-aligned p-well.

The pad is formed by a stacked metal 1 and metal 2 layer with a total thickness of 1.5 μm. Various pad openings are designed, the largest is 85 μm square. To improve adhesion of the metallization to the substrate, an additional polysilicon layer is integrated under the pad area. For sensors with overlapping pad and piezoresistive elements, the polysilicon was omitted. This results in a higher susceptibility for pad lift-off during the wire bonding. The test pad is connected to ground to shield electrical distortion. All sensor structures are fully CMOS process compatible. No additional post processing steps are needed for the ball bond sensors based on Al-pad structures.

The cut-off frequency of the sensor is mainly determined by the capacitance of the external wiring and the source resistance on the sense line. Sensor signal buffering electronics near or on the chip is hindered by the restricted space due to the bond head movement and by process temperatures of up to 200 °C. Amplification of the sensor signal is performed off-site to guarantee gain and offset stability. Therefore, the cut-off frequency has to be controlled by an appropriate source resistance of the sensor and R_{on} of the multiplexer circuitry. High source resistance is furthermore prone to capacitive coupling of noise on the test pad. On the other hand, self heating should be avoided.

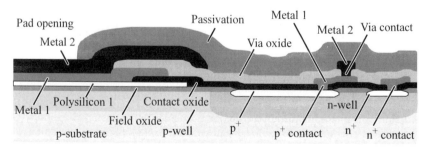

Fig. 2.21. Cross section of the CMOS process for the pad and the active sensor structures.

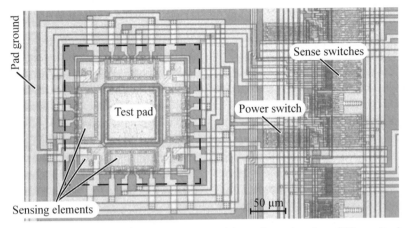

Fig. 2.22. Micrograph of a p$^+$ xyz-force sensor with a pad opening size of 65 μm. Dashed box marks the extension of the sensor that can be used for application specific sensor design without need of changing the connecting circuitry.

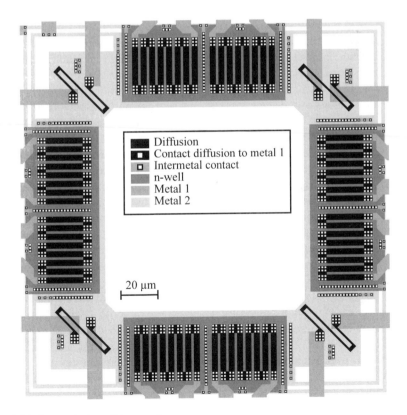

Fig. 2.23. Sensor design of the xyz-force sensor.

Figure 2.22 shows a micrograph of a xyz-force sensor with a pad opening of 65 µm. The sensor is connected to a multiplexer bus on the right. Interface connections between sensor structure and wiring enable adoption of the various sensor structures, without need of a redesign of the bus system. Test chips with optimized sensors for different applications are based on the same circuitry and chip dimensions. This standardized connecting framework facilitates the handling of the measurement system, as sensors can be changed without reteaching recipes of e.g. a wire bonder. The dimension of the sensor unit is 200.6 µm square. A table of the available sensor structures with their different pad designs is found in Sect. 3.1. The layout of a xyz-force sensor for a test pad opening of 85 µm is shown in Fig. 2.23.

2.3   Wedge Bond Sensor

The second bond formation of a thermosonic ball bonder significantly differs from the first bond formation for various reasons. The second bond is normally bonded to the substrate. In contrast to the first bond on the chip, there exists an extensive variety of substrate materials with different metallization stacks. As bond growth is heavily influenced by the used material combinations, different bond growth processes will occur for different materials and substrate contaminations.

Opposite to the ball bond with its rotational symmetry, the wire direction in case of the second bond breaks the rotational symmetry. Interplay between the reduced symmetry of the contact and the symmetry of the bonding system (ultrasound direction) can result in a strongly direction dependent process. Such processes are difficult to handle as bonding parameters have to be individually adjusted for the different wire directions. The performance of a wedge bond is not only determined by the bond strength, but also by a reliable tail formation. Inadequate tail bond strength can result in production stoppage or yield loss.

This higher complexity of the wedge bond results in a higher amount of bonding parameters, making it more susceptible to non-uniform bonding conditions. Bond process research with focus on the bondability of material combinations has to rely upon unaltered substrates. Fundamental research on the occurring bonding phases and bond parameter dependence has to focus on a representatively selected number of substrate and metallization families. In order to get a reliable and reproducible method for in situ and real-time inspection of the wedge bond process, our approach transfers the wedge bond process onto the die. Metallization stacks of the substrate can be reproduced by post-processing plating steps. However, stiffness and geometrical dimensions of the bonding substrate differ fundamentally between silicon and most of the leadframe substrates. Wedge bonds need larger pad openings due to the fact that the bonding zone and capillary center have an offset in wire direction. A too small pad will result in damage of the passivation ring (see Fig. 2.21) around the pad opening. To get reliable sensor signals, the capillary must not touch the passivation during the wire bonding. This is achieved by an enlarged pad opening in wire direction.

First measurements of wedge bonds were performed on sensor structures that were integrated beneath the pad [37]. Structuring of the sensor elements results in undesired pad surface height variations of up to 1.1 ± 0.3 µm. This difference was measured with a laser vibrometer and is to be related to the root mean square (rms) pad roughness of 0.5 ± 0.1 µm.

The measurement of the z-force based on σ_{xy} shear stress fields is sensitive to changes in the mechanical behavior of the test pad. This results in a high susceptibility to intrinsic stress changes and pad metallization thickness variations. The post-CMOS plating of thick metal layers on the test pad strengthens these effects. Buried sensor elements that are measuring σ_{zz} are less prone to changes in the surface conditions. As a maximum of two resistors of the Wheatstone bridge can be used as sensing elements, the sensor will inherently sense temperature differences between chip substrate and wire. The symmetry group of the σ_{zz} stress field is a subgroup of the symmetry of the thermal field of a heat sink. Lightly doped sensing elements have a lower TCR than highly doped resistors, and are thus beneficial if reference resistors have to be integrated. As the minimal spacing of such lightly doped elements is significantly larger, only one element of the Wheatstone bridge should then be integrated in the sensitive area to keep placement sensitivity small.

In conclusion, both the buried z-force sensor and the adjacent z-force sensor are sensitive to thermal effects. The buried sensor is sensitive to thermal gradients between the sensing and reference elements. The adjacent z-force sensor is sensitive to intrinsic stress fields, that are a function of the temperature.

The wedge bond sensor design splits up in a sensor for parallel and a sensor for perpendicular wire to ultrasound oscillation direction, as shown in Fig. 2.24. Even though the pad dimension of 75 µm by 134 µm is larger compared to that of the ball bond sensor, the size of the sensor unit remains unchanged. As a consequence, the wedge bond sensor is compatible to the ball bond connecting framework. The wedge bond sensor in Fig. 2.24a with wire direction in parallel to the ultrasound oscillation of the bonding tool, consists of slanting n^+ piezoresistive sensor elements integrated on the left and right sides of the test pad. The electrical dimension and electrical connection of the elements are equivalent to the n^+ xyz ball bond force sensor. In addition to this y-force sensor, the wedge bond sensor includes both

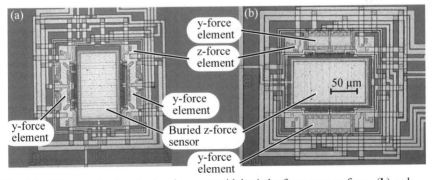

Fig. 2.24. Micrograph of wedge bond sensors with buried z-force sensors for x- (**b**) and y-direction (**a**) wires.

a buried z-force sensor and a z-force sensor identical to that used for the xyz-force sensor. From previous considerations about temperature stability follows that the buried z-force sensor has a better performance, especially for thick plated metal layers. The layout of the sensing elements is shown in Fig. 2.25 for the sensor for y-directed wires. The width and length of the buried z-force sensing and reference elements are adjusted to match the resistances. To get highest placement insensitivity, the sensing elements cover nearly the whole contact area. A σ_{zz} stress field component at the reference resistor region will diminish the sensor signal. In particular, this is the case for a capillary touching the surface at the end of the wedge bonding process. To avoid a distortion the reference resistor has to be integrated outside the pad area resulting in a significant increase of the design size.

The sensor for perpendicular wire to ultrasound oscillation direction is based on p^+ y-force sensing elements (see Fig. 2.24b). The z-force measurement arrangement is equivalent to wedge bond sensors with wire direction in parallel to the ultrasound oscillation. As a result of the sensors' difference in the y-force sensor designs, their y-force sensitivities are different.

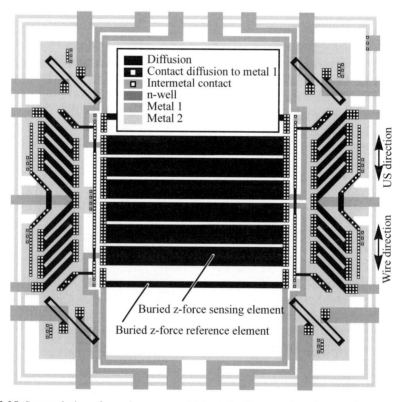

Fig. 2.25. Sensor design of a wedge sensor with buried z-force sensing elements for y-directed wires.

3 Measurement System

The xyz-force sensor is the key component of a measurement system optimized for the wire bonding applications. This chapter reports the main system features and elucidates the measurement signals.

3.1 Test Chip

Figure 3.1 shows a micrograph of a test chip with test pads for the wire bonding application. The die size is 3×3 mm^2. The core design of the test chip family is based on a standard 0.8 µm, double metal, single polysilicon CMOS process in order to ensure process portability.

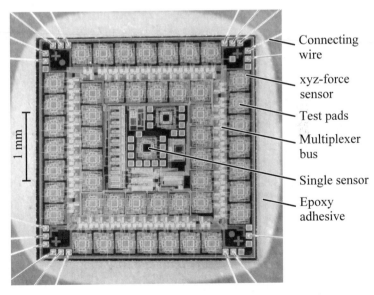

Fig. 3.1. Test chip with 48 xyz-force sensors connected to a multiplexed bus. Different test pads and sensor sizes are available on different test chips. The two additional single passivated sensors in the middle are added for special applications.

A multiplexer circuity selects one of the 48 xyz-force sensors to be measured. The 48 sensors are grouped in two concentric rows, an outer row consisting of 28 sensors and a staggered inner row of 20 sensors. The pad pitch within the rows is 300 μm, whereas the distance between the two rows is 550 μm. The corner sensors of the outer row are omitted to provide space for the connection pads. Multiplexer address, sensing, and power connection pads are placed in the edges of the die. These pads have to be connected with wire bonds prior to the measurement. The connection pad wires do not interfere with the wires bonded during the measurement. The connection pad naming is found in Fig. 3.2b. In case of the flip-chip test chip the signals are routed to the solder balls of the outer row. A set of test chips is available with exactly the same sensor and connection pad arrangement, but with various sizes and types of test pads and sensors as listed in Table 3.1. Thus, there is only one wire bonder recipe necessary for applications with different test chip types. For ball bond investigations, either passivated test pads (see Fig. 3.3a) or standard Al test pads (see Fig. 3.3b) are implemented. For flip-chip process and wedge bond measurements, additional metallization layers are deposited on the Al-pad. The schematic drawing in Fig. 3.3c shows a Ni-Au metallization stack on top of the test pad. A commercially available electroless plating process of PacTech, Nauen, Germany was selected that is able to support the large mechanical stresses during wire bonding and guarantees reproducible process test result. The nickel layer thickness is 5 μm. The thickness of the Au layer is 80 nm and 1 μm for the flip-chip application and wire bonding application, respectively. The thick gold layer used for the wire bonding application is electrolytically plated.

Figure 3.4 shows the cross-section of the sensor element and the solder ball of a flip-chip device. The two Al metal layers forming the pad are visible at the pad edge where they alternately overlap with oxide layers (see also Fig. 2.21). The polysilicon layer under the pad is also visible. This polysilicon layer is essential for improvement of the adhesion between the metal layers and the field oxide (refer to Sect. 2.2.1).

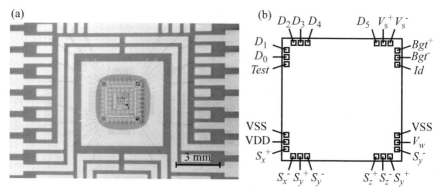

Fig. 3.2. Die (**a**) mounted on a custom made substrate. The pin-out (**b**) of the test chip is described in the text.

Table 3.1. Available test chip designs and their application.

Test chip name	Pad opening	Pad shape	Sensors	Application
$XYZ\text{-}Al85_{aligned}$ $XYZ\text{-}Au85_{aligned}$	$85 \times 85\ \mu m^2$	square	x, y, z	ball bond
$XYZ\text{-}Al65_{aligned}$ $XYZ\text{-}Au65_{aligned}$	$65 \times 65\ \mu m^2$	square	x, y, z	ball bond
$XYZ\text{-}Pass_{aligned}$	$85 \times 85\ \mu m^2$	square	x, y, z	bonder calibration and characterization
$XYZ\text{-}FC_{aligned}$	Ø 85 μm	octagon	x, y, z	flip-chip
$YZ\text{-}Al_{Wedge}$ $YZ\text{-}Au_{Wedge}$	$75 \times 134\ \mu m^2$	rectangle	y, z	wedge bond
$YZZ_b\text{-}Al_{Wedge}$ $YZZ_b\text{-}Au_{Wedge}$	$75 \times 134\ \mu m^2$	rectangle	y, z, buried z	wedge bond

The stress sensitive diffusion of the sensor is connected with contacts (see Fig. 3.4) to metal 1 on both ends. The sensor elements are covered with a metal 2 layer for protection against electromagnetic interference. To enhance the dc performance of the sensor under large temperature variations, these metal 2 layers should be omitted in future designs. Plastic flow in the metal layers can affect the offset stability of the sensor as demonstrated in Sect. 4.5.1.

The remaining space on the test chip is used for the multiplexer control circuitry, a temperature sensor, and single calibration sensors with passivated test pads as schematically shown in Fig. 3.3a. The digital multiplexer control circuitry and input protection are placed in the center of the chip. Additional circuitry detects open connecting wires and shorts between connecting wires. Every chip type has a mask-programmed identification byte (connection pads *Test* and *Id*).

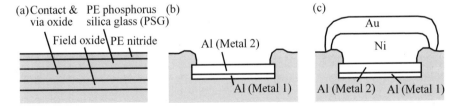

Fig. 3.3. Test pad surfaces (not to scale).
(a) Passivated test pad without any metallization.
(b) Al test pad (oxide and passivation layers are not drawn).
(c) Al-Ni-Au test pad (oxide and passivated layers are not drawn).

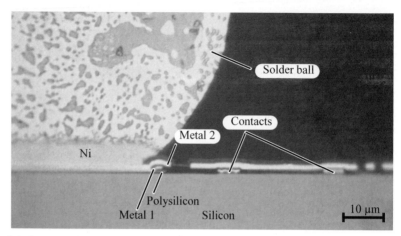

Fig. 3.4. Cross-section of the sensor element and the test pad structure.

The sensor to be measured is selected with a 6 bit pattern on the address bus D_0 to D_5. The power switches of the selected sensor are turned on, and the sensing line switches connect the sense lines of the sensor to the sense bus. As only one sensor at a time is connected to the bus, the power consumption and power bus size are minimized.

Figure 3.5 shows a simplified schematic of the on-chip and external circuitry. Only one of the three Wheatstone bridges of the sensor is shown (A). The digital circuitry on the chip and the sensors are powered by separate supply lines. The sensor voltage V_w can thus be adjusted within the range between 2 V and 5 V. All sensor signals are measured fully differentially with the signal lines S_x^+, S_x^-, S_y^+, S_y^-, S_z^+, and S_z^-. For exact sensitivity measurements, the voltage applied to the Wheatstone bridge can be measured differentially with the additional sensing lines V_s^+ and V_s^- common to all three Wheatstone bridges of the sensor. This allows for correction of the small voltage drop on the bus lines and switches.

During measurement the switches (B) and (C) and the p-FET (D) are activated with the enable signal. The on resistance (R_{on}) of the signal switches is 280 Ω. The signal conditioning is performed externally due to the high temperatures at the bonding site.

The sensor supply voltage is software selectable. The voltage is adjusted with a digital-to-analog converter (E) and buffered. This buffer (F) is short-circuit protected and supports a software selectable limit for the current (not shown in the schematic). The short-circuit protection is essential for the stepping cycle of the wire bonder as the downholder can touch the oven plate under absence of the leadframe.

The sensor signal $S^{+/-}$ and the applied voltage bus lines $V_s^{+/-}$ are connected to the external circuitry via bonding wires, the custom leadframe, and the custom downholder system. The differential signals are preamplified with an instrumental ampli-

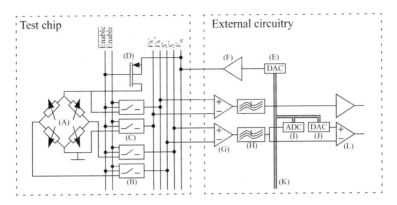

Fig. 3.5. Schematic of the multiplexer circuitry. Switching blocks are based on a combination of a p-FET and a n-FET. Enable lines are controlled by a demultiplexer circuitry. See text for explanation of components (A) to (L).

fier (G) and low-pass filtered (H). In a next step, an offset correction is performed on the sensor signal by subtracting a stored signal strength from the sensor signal. The subtraction is performed with the differential amplifier (L). The amplification stage also includes a software selectable variable gain amplifier. The offset value is recorded by the ADC (I) and sent to the DAC (J). The time point of the offset adjustment is software selectable and can be synchronized to either the first or the second trigger. All parameters of the electronic components are controlled by a Field Programmable Gate Array (FPGA) connected to the bus lines (K).

3.2   Wire Bonding Signals and Measurement System

Figure 3.6 shows raw microsensor signals of a ball bonded on a sensor $XYZ\text{-}Al85_{aligned}$. A typical recording time of the ball bonding sequence is 20 ms. This results in 100 000 measurement points at a sample rate of 5 MS/s. The black areas in the x- and y-force signal are due to the ultrasound oscillation of the bonding tool. An enlargement of the time scale shows the individual oscillations of the sensor signal in detail (d) for a time frame of 20 µs. Before the touchdown of the ball on the pad, the sensor signals are constant. If the hardware offset correction is activated and synchronized with the first trigger point, the sensor signal is zero due to the offset correction during the search time. The sensor signal before touchdown ultimately shows the noise floor of the measurement system.

After touchdown, the sensor signal is dominated by the forces acting between capillary and chip. These forces are either forces as a result of a relative movement between substrate and bond head or forces generated by the oscillation of the capillary at ultrasound frequency. The ultrasound force signal is especially dominant in the y-force signal as the y-direction is the ultrasonic excitation direction of the capillary vibration.

The different physical processes can be separated by selective filtering of the raw xyz-force signals, using three different digital filters. An infinite impulse response (IIR) low-pass filter is employed for extraction of the low-frequency part contained in the force signals. By forward-backward filtering, a zero-phase shift in the signal is achieved. In order to study the time evolution of the ultrasound force signals with its carrier frequency at 130 kHz, a finite impulse response (FIR) quadrature filter is used [1, 2, 3]. This filter technique allows for the extraction of the amplitude and phase information with optimal frequency and time resolution. Due to physical processes the force signal is amplitude modulated and also comprises higher harmonics that are amplitude modulated as well (see Chap. 5). All presented first and higher harmonic signals are based on this quadrature filter. The first harmonic is subsequently denoted fundamental as well. If the time evolution of the sensor signal is not of interest, a FFT (Fast Fourier Transform) filter based on flat-top or Kaiser window technique is employed for amplitude extraction [4].

As there is coupling between different bonding axes and physical processes, recording of the complete set of relevant data is essential for bond process investi-

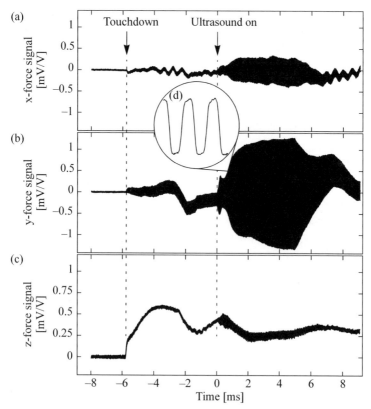

Fig. 3.6. Unprocessed microsensor signals of a real ball bond on Al-pads (**a**) to (**c**). Detail (**d**) shows the enlarged individual ultrasound wave oscillations. Measured on a WB 3088iP.

gation. The measurement system is designed for simultaneously recording microsensor signals and machine data. Table 3.2 lists the signals that are subsequently referred to. The two digital signals in the list are used for the triggering of the measurement. Only the timing information of the trigger points are saved of these two digital signals. The recording of the ball or wedge bond is selected by either triggering on the falling or rising edge of the initial point of the search process. Analog signals are sampled by the measurement system and are used for data analysis. The measurement system is capable of recording additional measurement signals.

In the following, a short overview on the different signals is given within the scope of a comparison between capillary imprint bonds and real bond process signals. A detailed treatment is found in Chaps. 4 and 5.

For particular applications it is interesting to measure the forces that act between capillary and chip while friction and deformation effects between the wire and pad surface are excluded. This is achieved by pressing the capillary directly on the passivation of the die (see Fig. 3.3a). An application of these *capillary imprints* is the calibration of the ultrasound system [5]. This method is furthermore extensively used for sensor characterization in Chap. 4 and machine characterization in Sect. 5.1. If the bonding tool (ceramic capillary) is in direct contact with a passivated sensor, the fundamental of the y-force is proportional to the ultrasound current of the transducer system, as shown in Fig. 3.7a. For all measurements, a small ultrasound current amplitude is present before the ultrasound bonding start (US on) at $t = 0$.

Table 3.2. List of signals recorded by the measurement system. μS : Microsensor, WB : Wire bonder.

Signal name	Type	Source	Description
x-force	analog	μS	sensor signal of x-direction
y-force	analog	μS	sensor signal of y-direction
z-force	analog	μS	sensor signal of z-direction
bridge voltage	analog	μS	voltage applied over Wheatstone bridge
start of search process	digital	WB	trigger point at search height
touchdown	digital	WB	trigger point of impact detection by the machine
bond force	analog	WB	bond force measured at the horn fixation
US current	analog	WB	ultrasound current of the transducer horn
z-position	analog	WB	z-position of the optical z-position measurement system

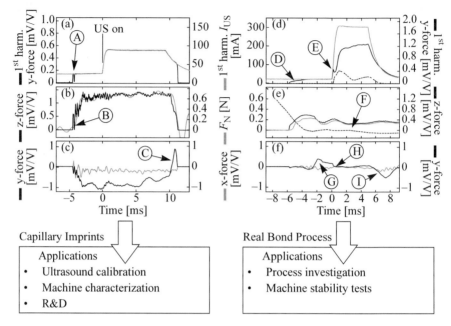

Fig. 3.7. Comparison of recorded signals between capillary imprints on passivated surface and real ball bonds on Al-pads. For explanation of the signals refer to the text. Measured on a WB 3088iP.

This minimal ultrasound current amplitude is needed to preserve the phase locked state of the used ultrasound system. At touchdown of the capillary (A), pulses that correlate with the z-force signal (B) indicate a bouncing of the capillary. This is not observable anywhere for real bonds due to the damping by the gold ball. The low-frequency z-force of the microsensor in Fig. 3.7b correlates well with the bond force of the machine except for high frequency oscillations in the z-force and a phase-shift in the bond force caused by analog filtering of the machine signal

The low-frequency x- and y-force as shown in Fig. 3.7c are solely accessible with microsensors. Coupling between the y-force and z-force causes a y-shift force. A normal force applied on the transducer horn results in a bending of the horn rod. The bending of the transducer results in a force in y-direction on the capillary tip. Due to friction between capillary and pad at the impact, the y-shift force shows a peak in opposite direction at capillary lift-off (C). Vibrational forces in x- and y-direction are observed under the condition of absent relative motion between capillary and chip. Sliding between the ceramic capillary and the chip passivation during the ultrasonic bonding time destroys the contact surfaces and thus needs to be avoided by using appropriate bond force and ultrasound settings. The high correlation between the machine parameters and the corresponding microsensor signals is the result of the rigid contact condition between capillary and chip surface.

In contrast to capillary imprints, the y-force of the real bond process does not correlate any longer with the ultrasound current in Fig. 3.7d. A slow increase in the fundamental of the y-force is observable during the impact phase (D). This increase is the result of a growing contact diameter between ball and pad and in particular a decrease of the ball height as demonstrated in Sect. 3.2.1. After ultrasound is turned on, a break-off point (E) marks the beginning of the friction process between ball and pad. Higher harmonics are observable, indicating a change in the wave form of the sensor signal. The dashed line in Fig. 3.7d exemplarily plots the 3rd harmonic with a vertical scaling of a factor three. The time evolution of the signal strength comprises information about the time evolution of the physical processes and thus indicates the progress of the bond forming process. A detailed discussion of the different processes and their implication in the bond growth and deformation is found in Sect. 5.2.

Beside the z-force signals in Fig. 3.7e, the z-position of the bond head (dashed line in Fig. 3.7e) is a good indicator for ball deformation if the contribution of the transducer system bending is eliminated. As the ultrasound tangential force increases during bond growth, the combination of normal force and ultrasound tangential force can induce ball deformation. The deformation is observable in the z-position signal at marker (F) in Fig. 3.7e. This second deformation process after the impact deformation is denoted *ultrasound enhanced deformation* and is discussed in Sect. 5.2. The low-frequency x- and y-forces in Fig. 3.7f are reduced due to friction and deformation processes. The y-force peak (G) is a force that results from the normal force transition from impact force down to the bond force. This force the result of the coupling between normal force and y-force (y-shift force). At the time (H), when ultrasound friction is induced, the low-frequency force drops to zero due to relative movement between ball and pad. The friction taking place at maximal and minimal displacement during the ultrasound oscillating cycle equalizes the mean value (low-pass signal) to zero. This implies that the maximal tangential force acting during the friction phase is defined by the friction force even if strong mechanical vibrations are superimposed. As bond growth progresses friction at the contact zone is stopped and vibrational forces (I) are acting on the contact zone.

The microsensor force signals are very susceptible to changes that affect the bonding process. This opens the door to fundamental process and machine characterizations. To examine a bond process it is often desired to explore the whole bond parameter space. This entails the recording of a large number of measurements. Consequently, the sensor has to be integrated in a system that is accessible for common process engineers with minimal training in the handling of the microsensors [6]. Reproducible measurements combined with a standardized data handling and a fast set-up time are the key for long-term measurements and machine characterization. The following paragraphs give a short overview on the custom made measurement system. Sensors with Al pads can only be used once due to the resulting bond. Even if many sensors are integrated on a test chip, the chip has to be replaced after some time. The custom made downholder (see Fig. 3.8) allows for fast chip exchange by just stepping the substrate with the test chips. Mechanical gold-plated

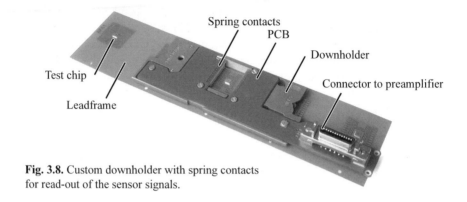

Fig. 3.8. Custom downholder with spring contacts for read-out of the sensor signals.

spring contacts assure high electrical connection reliability between the downholder and the substrate.

The requested temperature stability and the limited space is challenging the design of the contact tool. The whole wiring system with the spring contacts has to be integrated in the downholder plate with a maximal thickness of 4 mm. The system tolerates heater temperatures of 200 °C.

Real-time bonding without dead-times is targeted for examination of the wire bond process and the machine characteristics. High bonding speeds of the wire bonder demand short measurement cycles. High flexibility in the application requires the wire bonder to be the master, supplying trigger signals to the measurement system. To guarantee real-time functionality, the whole measurement flow control is implemented on a FPGA. For measurement cycles below 100 ms, dead-times of the PC system have to be handled to prevent unwanted measurement stops. During the measurement, status information of the PC is sent to the hardware tool. Therefore, absences of the PC can be detected by the FPGA. The missed wires are then marked in the database. This measurement flow control is prerequisite for long-term measurement series for wire bonder reliability tests.

The components of the signal circuitry are shown in Fig. 3.9 and Fig. 3.10. A small preamplifier close to the bonding place amplifies the sensor signals before they are processed in a custom made signal conditioning system that is controlled by the FPGA. Four identical fully differential amplifier stages are available on the preamplifier board. These four microsensor signal paths are routed together with the sensor power supply lines and additional measurement lines to the signal conditioning unit. The microsensor signals are offset corrected and amplified with an adjustable factor to get best signal recording resolution for all signals. All components of the signal conditioning units are controlled by a FPGA. Analog machine signals and trigger signals from the wire bonder are also fed through the signal conditioning unit to suppress electrical noise. The FPGA also addresses the multiplexer of the sensor chip. All settings of the measurement system are software selectable and are stored together with the measurement data. Figure 3.11 schematically shows the timing synchronization between the measurement signals and the machine timing. The measurement zero point is attached to the touchdown detection of the machine.

The touchdown detection of machine is delayed by two reasons. To achieve a reliable touchdown detection, the bond force has to cross a defined force level. Analog filtering of the force signal results in a phase shift. Based on the fundamental of the y-force signal and the z-force of the microsensor a precise touchdown time determination is achieved. This is due to the fact that digital acausal filtering is used offering zero phase delay.

The start of the search process is the transition from rapid bond head movement to a z-position approach in direction of the pad with a defined velocity. The search process duration is an important machine parameter for bonding speed optimization. A too large search time results in a lower bonding speed, whereas a too small search time bears the risk of an 'early touchdown', i.e. the ball touches the pad before the start of the search process. This is due to the fact that the z-position of the current bond is a priori unknown and has to be estimated from the z-position of earlier bonded balls. The final z-position of the bond head is a function of the die tilt and ball height. A longer search time generally reduces the vibration amplitudes during the ball bonding process.

For the triggering of the microsensor measurement a dual trigger configuration is used. The first trigger signal is the start point of the search process. At this trigger point the multiplexer is switched to the next pad. A subsequent offset correction is

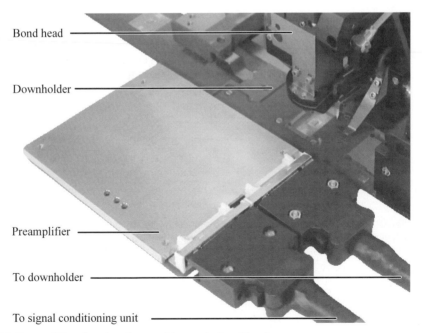

Fig. 3.9. Preamplifier placed at close position to the bond head.

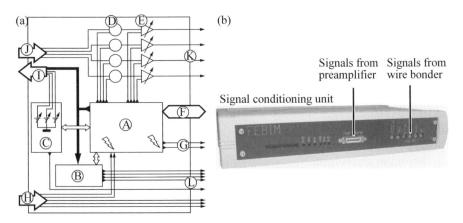

Fig. 3.10. Schematic (**a**) and micrograph (**b**) of the signal conditioning unit. FPGA (A), cross-point switch (B), sensor source (C), offset correction (D), variable gain amplifier (E), digital interface to PC (F), trigger lines (G), signals from wire bonder (H), power and multiplexer address signals to sensor (I), sensor signal from preamplifier (J), analog signals to recording cards of PC (K and L).

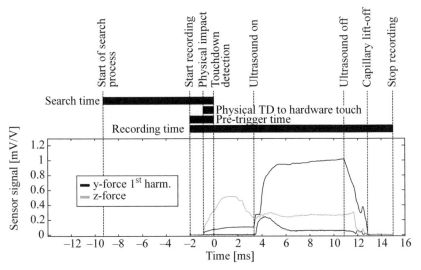

Fig. 3.11. Synchronization between wire cycle and measurement flow. The measurement is synchronized with the second trigger issued at touchdown detection of the machine.

performed if the offset correction is synchronized to the first trigger. The second trigger point is the touchdown detection of the machine. This time point is used for the data recording. The start of the search process is not suited for the data recording, as the search time can vary for sequentially bonded balls. To record the physical touchdown, the measurement is started in a pre-trigger mode, i.e. the recording is started a fixed time before the receipt of the trigger by using a circular sampling data buffer.

The first and second trigger times are recorded by the FPGA and used for complete reconstruction of the time information. In addition to these two trigger signals issued by the wire bonder, the software offers time synchronization with distinguished points of the measurement signal. This includes routines for determination of the physical touchdown in the microsensor y-force and z-force and the US on point. Based on these timing points further distinguished time marks are determined in the signals. Based on the time marks signal parameters can be extracted for comprehensive signal analysis.

3.2.1   Ultrasonic System

Thermosonic wire bonding systems use ultrasound oscillations to enhance the bonding process. The ultrasound is induced by transversal oscillations of the bonding tool (see Sect. 1.2). The amount of ultrasound energy that is applied to the bonding zone has to be controlled tightly to get reproducible results. Various calibration methods are used [5]. The closer the calibration point is to the bonding zone the better is the control of the ultrasound energy applied to it.

Under normal condition the transducer system is driven in constant current mode. The ultrasound controller readjusts the transducer driving voltage to regulate the transducer current to follow a predefined profile. The applied current on the transducer is a very good measure of the displacement at the horn front.

For the study of the bonding mechanism, the knowledge of the force and displacement at the bonding zone is essential. Due to the small dimension, the ball is hardly accessible for conventional measurement methods. The absence of appropriate measurements methods also explains the fact that ultrasound is always adjusted to values that are related to the ultrasound amplitude or transducer amplitude but not to the physical conditions at the capillary tip. The presented force microsensors are up to now the only systems that in situ and in real-time measure the force that is acting on the contact zone. To relate the forces measured by the microsensor at the contact zone to the amplitude at the transducer horn front, the properties of the mechanical system of the ceramic capillary have to be considered.

Capillary Model. By using the symmetry of rotation of the capillary, the displacement $\psi(z, t)$ of the capillary is described by the differential equation [7]

$$\partial_z^2 EI\partial_z^2\psi + F_z\partial_z^2\psi + c_{\text{cap}}\partial_t\psi + \rho A\partial_t^2\psi + \Phi_{\text{Timoshenko}}\psi = 0 . \tag{3.1}$$

The first term of the fourth-order differential equation stands for the bending stiffness, defined by the elastic modulus $E(z)$ and the surface moment of inertia $I(z)$ [7]. The surface moment of inertia is calculated by an integration over the cross-sectional area of the capillary

$$I(z) = \int_{Area} r^2 df \qquad (3.2)$$

at a given z-position. The prestress of the capillary is described by the second term for an applied normal force F_z. This prestress term is of minor influence if the excitation frequency does not coincide with the natural frequency of the capillary vibration. This condition is satisfied for the used bonding system driven at a frequency of 130 kHz, as the fundamental natural frequencies of the capillary are typically at 70 kHz, 300 kHz, ... The excitation frequency is defined by the transducer horn.

A viscous damping term with damping coefficient c_{cap} is added to improve the stability of the solution. An exact determination of the damping constant is difficult. An upper limit can be estimated from recorded profiles of the capillary displacement and phase, as viscous damping will change the phase at the node. The inertia term includes the cross-section area of the capillary $A(z)$ and the density ρ of the capillary. The material or boundary condition parameters used for the simulation are shown in Fig. 3.12. The last term in equation (3.1) is the Timoshenko correction accounting for rotatory inertia and shear corrections [7]. Based on a similar differential equation the oscillation behavior of wedge tools is studied in [8, 9]. As the capillary displacement at the transducer horn fixation has no more rotational symmetry, the upper boundary conditions are applied below the transducer horn. The two upper complex boundary conditions are defined by the displacement excitation

$$\psi(0, t) = \psi_0(t) \qquad (3.3)$$

and a rotational stiffness

$$EI\partial_z^2\psi(0, t) = K_{TH}\partial_z\psi(0, t) \qquad (3.4)$$

of the horn. The excitation displacement $\psi_0(t)$ is a harmonic function defined by the amplitude A_H and the excitation frequency ω. The phase is set to zero. The

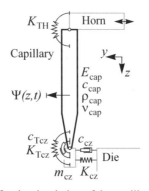

Fig. 3.12. Mechanical model for the simulation of the capillary oscillation.

oscillation amplitude and the rotational stiffness have to be determined from laser vibrometer scans of the capillary oscillation amplitude along the z-axis (see Fig. 3.13).

The remaining two boundary conditions are assigned at the capillary tip. For a free oscillating tip, the bending moment $\partial_z^2 \psi(l, t)$ and bending force $\partial_z^3 \psi(l, t)$ must vanish at the end of the capillary ($z = l$). If the capillary tip is in contact with the passivation, the force and moments are related to the displacement and tilting [10]. The contact stiffness K_{cz}, the lumped mass m_{cz} (silicon underneath the contact), and the viscous damping constant c_{cz} are the parameters of the force boundary condition

$$\partial_z EI \partial_z^2 \psi(l, t) = K_{cz} \psi(l, t) + c_{cz} \partial_t \psi(l, t) + m_{cz} \partial_t^2 \psi(l, t). \tag{3.5}$$

The stiffness of the contact is determined from the analytical calculations or FEM simulations in Sect. 2.1.3. The damping at the contact will significantly influence the amplitude at the contact. As laser vibrometer measurements of the clamped capillary tip amplitude are extremely difficult, an exact determination of the damping conditions at the contact is not possible. However, the force acting on the contact is in first approximation independent of the damping conditions. This force value is measured by the microsensor. The force acting on the contact is an important process parameter as it is related to ball deformation, as explained in Sect. 5.2.3.

The rotatory stiffness of the contact fulfills

$$-EI \partial_z^2 \psi(l, t) = K_{Tcz} \partial_z \psi(l, t) + c_{Tcz} \partial_t \partial_z \psi(l, t). \tag{3.6}$$

The mass of the silicon is approximated by a lumped mass that is calculated from a harmonic FEM simulation. It is estimated from the amplitude response at the capillary tip as a function of the frequency for quasi-static conditions (excitation frequency below first natural frequency of the capillary). Due to the large stiffness of the silicon, the moved mass is of minor influence. The gold ball is also included into the model with the mechanical properties of the gold being approximated by a linear elasticity with $E_{Au} = 79$ GPa and $\nu_{Au} = 0.42$ (see Sect. 5.2).

The differential equation is solved with an one-dimensional FEM solver for a harmonic displacement described by

$$\psi(z, t) = \Phi(z)e^{i\omega t}. \tag{3.7}$$

The time independent displacement function $\Phi(z)$ is a complex function of the z-position, the capillary geometry, the capillary material parameters, and the boundary conditions. The availability of the capillary material properties is most critical. The density and the geometry are determined by weighting and cross-section preparation. The Young's modulus is known exactly only for alumina (Al_2O_3) ceramic materials. Simulations and measurement of this capillary type correlate well. The plot in Fig. 3.13 shows simulation results of a free and clamped capillary excited at $\omega = 2\pi \cdot 130 \text{ kHz}$. The vibration shape of the free solution is verified with a laser vibrometer measurement. The displacement amplitude of the capillary in the transducer horn can be measured but is not described by the simulation as the transducer horn is not included in the simulation model. The excitation amplitude $A_H = 1442 \text{ nm}$ results in a free tip amplitude $A_T = 2090 \text{ nm}$ and a clamped force amplitude at the capillary tip of $F_y = 357 \text{ mN}$. As a change in the Young's modulus directly affects the force at the tip, the Young's modulus of other capillary materials had to be used as fit parameter.

Damping influences can be examined in the phase relation between excitation (horn displacement) and capillary displacement $\psi(z, t)$ near the node of the vibration. Figure 3.14 shows simulation results of the phase and displacement for different viscous damping coefficients.

From comparison with measurements a damping coefficient of $c_{\text{cap}} = 5 \text{ kg m}^{-1}\text{s}^{-1}$ fits well with the simulation. Away from the resonance frequency, the bending stiffness (first term) and inertia (fourth term) in the differential equation (3.1) are the dominant terms. As the surface moment of inertia has a functional dependence like

$$I \propto r^4, \tag{3.8}$$

capillary radius constrictions (as e.g. at the capillary tip) will significantly alter the force that can be applied to the contact zone.

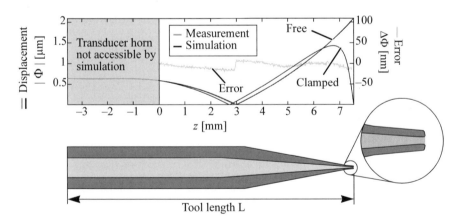

Fig. 3.13. Absolute value of the displacement of a free and clamped capillary SBNE-38BD-C-1/16-XL-20MTA as obtained by laser vibrometer measurements and by solving Eqn. (3.1). Measurement and simulation are on top of each other with a deviation of less than 20 nm as indicated with the error plot. The error $\Delta\Phi$ is the difference between the simulated and measured amplitude.

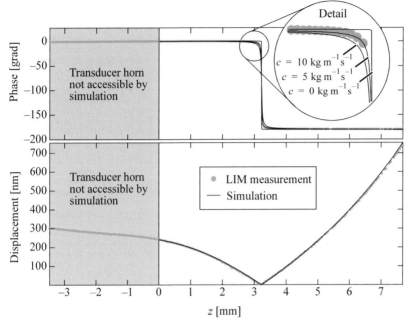

Fig. 3.14. Phase and displacement amplitude of a free capillary as obtained by simulation with different damping coefficients in comparison to a laser vibrometer measurement (LIM).

As the model is linear, nonlinear boundary conditions, as occurring e.g. during the friction process, are excluded. The contact between capillary and transducer horn is only approximated. An improved model would demand a three-dimensional model, as the transducer horn breaks the rotational symmetry. This would request the mechanical simulation of the whole transducer system. As the interest of this simulation model is put on the capillary-ball contact, the simulation of the full system would be inadequate for parametric simulations due to the large calculation load.

Ball Height Sensitivity. It is observed that the fundamental of the y-force correlates with the ball height even in the absence of wave form modulating effects. To examine this behavior, gold balls with different volumes are pressed on a passivated microsensor test pad. A volume change in the free air ball result mainly in a change of the ball height. A constant ultrasound amplitude below the friction limit is applied. The free air ball diameter is varied in the range of 41 μm and 65 μm which results in ball heights of 5.5 μm to 33 μm. Impact and bond force are identical. Two measurement series with impact forces of 700 mN and 800 mN are recorded. A small ultrasound amplitude is selected to inhibit friction and deformation effects. As the pad is passivated, the ball sticks in the capillary after the ball bond process. The

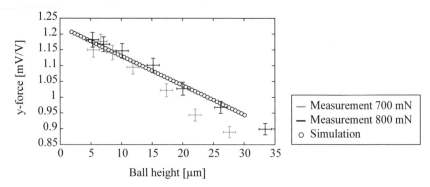

Fig. 3.15. y-force sensor signal strength as function of the ball height. The simulated dependence is a result taken from the capillary model.

contact zone diameter can thus be measured with an optical microscope. The ball height is measured with a scanning electron microscope (SEM). The measured y-force amplitudes are plotted as a function of the measured ball height in Fig. 3.15. The simulation result correlates well with the measurement.

The decrease of the y-force amplitude for increasing ball heights is observed for real bonds as well. Variations in the y-force amplitude of fully bonded balls is therefore an indicator for ball height variations.

4   Characterization

Each sensor experiences performance limitations due to physical constraints, cross-coupling of physical effects, process variations, and design trade-offs. The impact of the particular non-idealities depends on the specific sensor application. Table 4.1 lists some sensor properties and their significance for the applications wire bonding and flip-chip bonding.

In the following chapter on sensor characterization, sensor calibration, sensitivity variations, signal-to-noise ratio, and the offset are discussed. If the design name is not explicitly mentioned, the presented data is referred to the xyz-force sensors $XYZ\text{-}Pass_{aligned}$ or $XYZ_{aligned}$. A summary of the sensor properties is found at the end of this chapter.

Table 4.1. Significance of sensor properties for the two applications wire bonding and flip-chip bonding.

Property	Wire bonding	Flip-chip
Linearity	important	important
Offset	less important	less important
TC of offset	less important	crucial
Sensitivity stability	important	crucial
Sensor size	important	important
Dynamic range	important	less important
S/N ratio	less important	less important
Mechanical stability	important	less important
Packaging process compatibility	important	important

4.1   General Data

The room temperature resistances of the p^+ and the n^+ piezoresistor elements are $2900 \pm 50\ \Omega$ and $1950 \pm 50\ \Omega$, respectively. The resistivity slightly depends on the die position on the wafer. The parallel connection of the Wheatstone bridges for all three axes results in a total power consumption of 30 mW (6 mA @ 5 V), even though only one sensor is active at a time. The chip temperature increase due to the self-heating is 0.65 K at room temperature. The temperature increase is measured with a temperature sensor integrated on a chip that is bonded on a substrate with epoxy adhesive. This temperature sensors is based on a bandgap reference that yields a PTAT (proportional to absolute temperature) voltage. The substrate is clamped on the heater plate of the wire bonder. The heater plate thus acts as heat sink. The bridge voltage measurement accuracy is ±1 %, with systematical errors of the wiring resistance being included in this error.

4.2   Sensor Calibration

The sensors used in this work are inherently insensitive to stress fields that are constant over the sensor region. Therefore, commonly used calibration methods, as e.g. bending bridge [1] and beam bending methods [2], are not applicable. In order to calibrate the sensor directly, defined forces have to be applied at the contact zone. Due to the small dimensions, the applicable techniques are limited. The calibration approaches used here are listed in Fig. 4.1. The use of the flip-chip device and the shear tester directly yields the sensitivity. The use of the ultrasound system incorporates the capillary model, as the ultrasound force at the tip has to be determined. The simulations of the sensor signal strength are based on stress field calculations and are prone to inaccuracies due to model simplifications and errors in the material parameters. In the following the employed calibration methods are discussed individually.

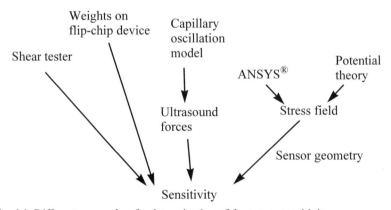

Fig. 4.1. Different approaches for determination of the sensor sensitivity.

(a)   (b)

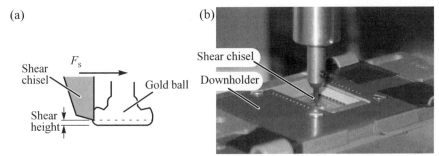

Fig. 4.2. Schematic drawing of the ball shear test (**a**). Shear tool positioned close to a ball of the chip (**b**).

4.2.1   Calibration with Shear Tester

Shear force measurements were performed with a Dage 4000 shear tester of Dage Precision Industries, Aylesbury, United Kingdom. The bond strength of a ball bond is commonly characterized by measuring the force that is needed to shear the ball as schematically shown in Fig. 4.2a. The measured force value together with the failure mode is a widely accepted method of determining the quality of a ball bond [3]. During production, samples are tested off-line to monitor the bond quality.

For sensor calibration, the microsensor signal and measured shear force of a nondestructive test are synchronously recorded. This calibration method can only be used with bonded wires or solder balls on the pad. Fig. 4.2b shows the shear chisel positioned above a chip. The custom made downholder system was used for electrical connection of the chip. Calibration is restricted to substrate temperatures lower than about 80 °C, because the shear chisel is a heat sink and causes temperature gradients on the chip. Furthermore, a temperature gradient on the shear tool diminishes the accuracy of the test system. The measured sensitivity of the x-force and y-force sensors is $g_{yy} = 10.3 \pm 0.5$ mV/V/N at $T_s = 25$ °C and $g_{yy} = 9.9 \pm 0.5$ mV/V/N at $T_s = 80 \pm 5$ °C for solder balls.

4.2.2   Calibration with Gauge Weights

The main difficulty of sensor calibration is due to the small dimension of the contact zone where the force has to be applied to. This problem can be circumvented by using a flip-chip bonded die for sensor calibration. In this case, the shear forces and the normal force are applied to the whole chip. As the to be measured contacts are the only mechanical link to the substrate, the recording of all sensor signals yields a calibration value. By assuming equal sensitivity of all pads the sensor sensitivity is given by the sum of the sensor signals of all solder ball connections divided by the applied force. Figures 4.3a and (b) show the measurement result and the measurement setup, respectively. Calibration weights between 5 g and 250 g are used. At elevated temperatures, this method is limited by offset drifts. Not exactly centered

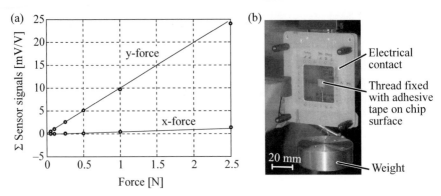

Fig. 4.3. Total microsensor signal strength (**a**) as function of the attached weight. The total microsensor signal is the sum of the signal from the 28 sensors with attached solder balls. Measurement setup (**b**) with calibration weight fixed on the flip-chip test die.

and aligned force exertion results in signal components in perpendicular direction (x-force in Fig. 4.3). Having 28 soldered interconnects between chip and substrate, the average y-force sensitivity is $g_{yy} = 10.1$ mV/V/N.

The calibration of the z-force sensor is performed with a triaxial translation stage of Physik Instrumente (PI) GmbH, Karlsruhe, Germany, to position a ceramic capillary on the passivated sensor pad. A defined force is applied on the capillary using a linear bearing. To exclude friction influences of the bearing, the applied force is simultaneously measured with a commercially available force sensor. The calibration of the z-force sensor yields a sensitivity $g_{zz} = 2.24 \pm 0.05$ mV/V/N at $T_s = 25$ °C.

4.2.3   Calibration with Ultrasound Force Measurements

The use of a wire bonder for sensor characterization has some distinguished advantages with respect to manually positioned probing systems given by a high positioning accuracy and high repeatability. By measuring forces at ultrasound frequency, offset drift and 1/f noise can be removed by digital filtering. During capillary imprints, the rigid ceramic of the capillary tip is pressed directly on the passivation of the sensor. The tip imprint of ball bonding capillaries is ring shaped. Therefore, the extension of the contact zone is not exactly known, and the contact pressure is generally not equally distributed over the whole contact zone. The main supporting point of the capillary may therefore be off-centered even thought the capillary is centered. To correct the resulting uncertainty in the determination of the ultrasound force value, the sensor signal is recorded on a grid of bonding positions as shown in Fig. 4.4a. This results in a sensitivity surface as shown in Fig. 4.4b. The placement response of the sensitivity is equivalent to a saddle surface. The surface shape can be used for extraction of a precise value (saddle point), even if the supporting point

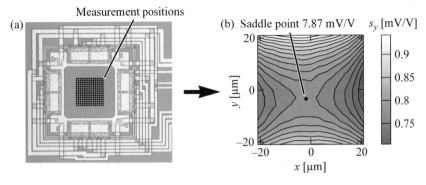

Fig. 4.4. Y-force sensor response recorded while bonding at grid positions (**a**) yields a sensitivity versus placement graph (**b**). The sensitivity at the saddle point is 7.87 mV/V. The black circles in (**a**) mark the different measurement positions with a pitch of 3 μm.

of the capillary is not accurately known. This procedure is e.g. used for the ultrasound calibration method reported in [4].

In order to extract a sensor sensitivity from ultrasound amplitude measurements, the force at the capillary tip has to be known. The capillary oscillation model of Sect. 3.2 is used to relate the transducer horn amplitude to the force at the tip. To achieve a reliable estimation of the force at the tip, laser interferometer displacement measurements were performed to record the displacement profile of the freely oscillating capillary. Displacement and bending of the capillary just below the horn fixation are extracted from the recorded displacement profile and used as input parameters for the upper boundary conditions.

To get a well defined contact condition for the sensor characterization task, a ceramic imprint capillary is used. The designation of the Alumina (Al_2O_3) capillary is TAMP-C-.001"-XL, produced by SPT Roth Ltd, Lyss, Switzerland. The tip of the capillary is flat and measures 25 μm in diameter. A cross-section of the capillary tip is shown in Fig. 4.5b. The length of the capillary beneath the fixation at the horn is 7650 ± 50 μm. The weight of the ceramic capillary is 58.83 ± 0.1 mg. In combination with a volume calculation based on the cross- section geometry of the capillary, the weight measurement yields a density of 3760 ± 75 kg m^{-3}. This value is slightly lower than the value for alumina (Al_2O_3) 99.9 % as listed in Table 4.2. The measured geometry is moreover an input parameter for the model.

Beside the geometry of the capillary, the mechanical properties Young's modulus and Poisson ratio directly affect the stiffness of the capillary and, thus, the force that is transferred to the bonding zone. The Young's modulus of alumina ceramic depends on admixtures, manufacturing process, and porosity of the polycrystalline structure that result in a relatively large variance. To account for the porosity of the ceramic injection molding process that is used for capillary production, the Young's modulus is selected from tabulated values by matching the measured density. To estimate the possible error due to uncertainties in the material constants, the

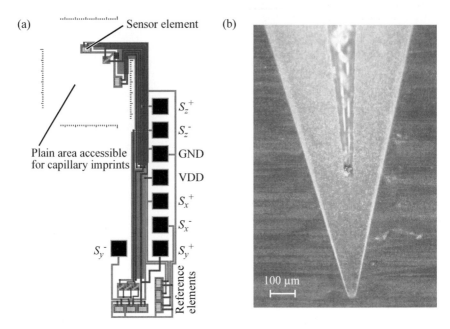

Fig. 4.5. Sensor structure (**a**) used for measurement of stress fields and cross-section of used ceramic imprint capillary tip (**b**).

Young's modulus was used as fit parameter. Satisfactory matching with the laser interferometer measurements was found for values $E = 320 \pm 40$ GPa. Figure 4.6 shows a profile of a laser interferometer measurement compared with the model of a freely oscillating capillary. In contrast to other capillary types of the same material, the matching is moderate. A possible explanation is a lower density of the ceramic material at the tip of the capillary. Inspection of the capillary cross-section indicated higher porosity at the tip.

Table 4.2. Material parameters of used capillaries. Values used for simulation are underlined.

Capillary material	Property	Value at 25 °C	Ref.
Alumina Al_2O_3 99.9 %	Density	3990 kg m^{-3} 3760 kg m^{-3}	[5] measured
	Young's modulus	298 … 388 GPa 309 GPa [a]	[6] [7]
	Poisson ratio	0.23	[6]

(a) Sintered ceramic with density of 3710 kg m^{-3}

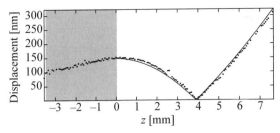

Fig. 4.6. Simulation (solid line) and measurement of the displacement of the freely oscillating capillary TAMP-C-.001"-XL. The shaded area marks the part of the capillary that is clamped by the transducer horn.

The amplitude of the freely oscillating capillary measured at the capillary shaft just underneath the horn is 150 ± 5 nm. By changing the lower boundary from the free condition to the boundary condition of a capillary-chip contact, the simulated tangential force amplitude acting on the contact zone is 82 mN. The most important error contributions are the uncertainties in the Young's modulus (± 9 mN), tip diameter (± 4 mN), and bending stiffness of the horn (± 2.5 mN).

The saddle point value is determined with a 2-dimensional pure quadratic fit of the y-force sensor signal strength surface. The saddle point value $s_y = 0.780 \pm 0.005$ mV/V is the average of the calculated saddle point value of 5 measurement runs. The laser interferometer displacement profile is recorded at the wire bonding machine to get identical measurement conditions for laser interferometer measurements and sensor measurements. The normalization of the sensor signal by the calculated tangential y-force at the tip yields a sensor sensitivity $g_{yy} = 9.5 \pm 1.1$ mV/V/N.

To investigate the stress field caused by a local contact, a special sensor was designed with only one sensor element near the contact. A plain area around the sensor is accessible for capillary imprints. The surface around the sensor elements is scanned with capillary imprints. The recorded sensor signal strength at the different positions gives information about the spatial extension of the stress field. Figure 4.5a shows the sensor design with p$^+$ sensor elements aligned along [110], [1$\bar{1}$0], and [010]. Each of these sensor elements is completed to a full Wheatstone bridge with three identical sensor elements placed 800 μm distant to the measurement area.

Figures 4.7a and (b) show the signal strength recorded by a p$^+$ sensor aligned in [1$\bar{1}$0] direction. The spacing between the scanning points is 20 μm. For one measurement the signal strength of up to 381 positions is recorded. The hatched area marks the region that is non-accessible for the measurement as an imprint there would affect the layer stacks containing metal and destroy the structures. A high contact force of 500 mN is chosen. This results in surface aberrations that are visible with light microscopy. Due to abrasion of the edges, the effective contact zone radius $r_c = 8 \pm 2$ μm is smaller than the original tip radius of 12.5 μm. This can be attributed to abrasion of the capillary that breaks the tip edges as a result of the large stress fields.

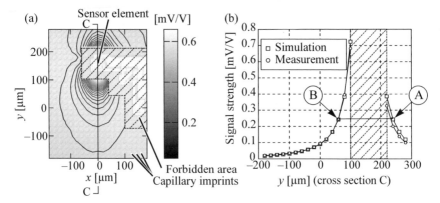

Fig. 4.7. Signal strength of the p$^+$ y-force sensor element as function of the position of the point contact (**a**) and as profile at x = 0 (**b**). Points in the hatched areas were not measured to prevent any damage to the sensor elements. The micrograph in the background shows the sensor elements and capillary imprint traces that are the result of the high normal contact pressure.

The horn amplitude and the displacement profile of the freely oscillating capillary (TAMP-C-.001"-XL) are measured with laser interferometry and used as input for the capillary tip force calculation with the capillary simulation tool. The resulting shear force applied to the contact is 82 ± 10 mN. The sensor signal strength is a function of the position of the capillary imprint for the above contact force. The result of a sensor signal simulation for above shear force value is compared to measurement data in Fig. 4.7b. The center of the sensor element is at position $y = 150$ μm. The measurement shows a significantly lower signal strength for measurement points above the sensor structure (A) than for positions at same distance below the sensor structure (B). The sensor is connected with aluminum lines above the structure. This symmetry violation could be a possible explanation for the lower sensitivity to forces applied above the sensor.

Table 4.3 summarizes the x- and y-force calibration values for sensor $XYZ_{aligned}$ and compares the experimental values with simulation results. The measured values are generally lower than the simulation values. The simulation model including the anisotropy of silicon and the Si_3N_4 / SiO_2 layers (see Sect. 2.2) is closest to the experimentally determined calibration values. The comparison of the simulation results with the stress field measurements indicate that a spatially correct description of the sensor performance can be achieved by approximating the sensor structure with anisotropic silicon. However, all parameters that are connected to the absolute value of the sensitivity are generally overestimated by simulations based on bulk silicon alone. The layers of the CMOS process have to be included in the model to get a precise estimation of the sensitivity.

Table 4.4 summarizes the z-force calibration values for sensor $XYZ_{aligned}$ and compares the experimental values with results of simulations.

Table 4.3. Comparison of x- and y-force sensitivities of sensor $XYZ_{aligned}$ with pad opening of 85 µm.

Method	Sensitivity [mV/V/N] at 25 °C	Error [mV/V/N]
Shear tester (ball bond on Al pad, $r_c = 21$ µm)	11.5	1 [a]
Shear tester (solder balls on Ni-Au coated Al pad, $r_c = 42.5$ µm)	10.32	0.5
Calibration weights (solder balls on Ni-Au coated Al pad, $r_c = 42.5$ µm)	10.1	0.5
Ultrasound ($r_c = 12.5$ µm, XYZ-$Pass_{aligned}$)	9.5	1.1
Simulation (isotropic, homogenous Si, stress field calculated with ANSYS, $r_c = 12.5$ µm)	14.51	—
Simulation (isotropic Si, analytical calculation of stress field, $r_c = 12.5$ µm)	14.48	—
Simulation (anisotropic Si, stress field calculated with ANSYS, $r_c = 12.5$ µm)	15.42	—
Simulation (anisotropic Si, sensor coated with Si_3N_4 and SiO_2 layers, stress field calculated with ANSYS, $r_c = 12.5$ µm)	12.28	—
Simulation (anisotropic Si, sensor coated with Si_3N_4 and SiO_2 layers, stress field calculated with ANSYS, $r_c = 21$ µm)	13.08	—
Simulation (anisotropic Si, sensor coated with Si_3N_4 and SiO_2 layers, stress field calculated with ANSYS, $r_c = 42.5$ µm)	14.18	—

(a) Measurement based on uncalibrated measurement system.

Table 4.4. Comparison of z-force sensitivity of sensor $XYZ_{aligned}$.

Method	Sensitivity [mV/V/N] at 25 °C	Error [mV/V/N]
Capillary imprint with translation stage (XYZ-$Pass_{aligned}$)	2.24	0.05
Simulation (anisotropic Si, sensor coated with Si_3N_4 and SiO_2 layers, stress field calculated with ANSYS, $r_c = 42.5$ µm)	1.25	—
Simulation (anisotropic Si, sensor coated with Si_3N_4 and SiO_2 layers, stress field calculated with ANSYS, $r_c = 21$ µm)	1.76	—
Simulation (anisotropic Si, sensor coated with Si_3N_4 and SiO_2 layers, stress field calculated with ANSYS, $r_c = 12.5$ µm)	1.85	—

4.2.4   Noise

A large signal range not only demands for a high sensitivity but also requires a low noise floor. Fig. 4.8 shows the noise floor of the x- and y-force sensors the and measurement system. The white noise level is dominated by the external preamplifier stage with $11 \, nV/\sqrt{Hz}$. The white noise level of the system corresponds to a noise equivalent force (NEF) of $380 \, nN/\sqrt{Hz}$ for a 5 V Wheatstone bridge excitation. The noise between 0.1 Hz and 10 Hz is 1.3 μV_{pp}, equivalent to 26 μN. The coupling of noise from the environment is minimized due to fully differential sensor signal lines. The above noise characterization was performed externally from the bonding machine. If the sensor system is used on a wire bonder platform, the electrical connections between the measurement system and wire bonder and electromagnetic AC fields induce additional distortions. These distortions manifest themselves in spikes in the noise spectrum. Therefore, appropriate signal filtering can significantly increase the signal range for e.g. ultrasound force measurements.

4.2.5   Sensor to Sensor Variations

For characterization of measurement to measurement deviations, 50 uniformly spaced dies were taken from a wafer. The passivated sensor XYZ-$Pass_{aligned}$ is selected for measurement. The sensitivity measurements are performed at ultrasound frequency to exclude any offset related deviations. All measurements are performed with a ceramic imprint capillary (TAMP-C-.001"-XL) with a tip diameter of 25 μm. Impact force and bond force are both 400 mN. The average y-sensor signal strength is 0.927 mV/V. No significant sensitivity drift is observable over the wafer. Table 4.5 summarizes the sensitivity deviation of the single passivated sensor placed in the center of each die.

The measurements on one wafer without saddle point characterization uses the optical alignment system for proper sensor to capillary alignment. Due to small tilts of the die, the real contact position may vary from sensor to sensor even if the opti-

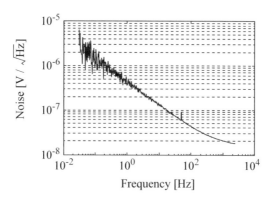

Fig. 4.8. Noise of the sensor and measurement system at ambient temperature and 5 V source voltage.

Table 4.5. Variation of the sensor sensitivity.

Variation of sensitivity	Standard deviation [μV/V] of 50 samples	Normalized [%]
Single sensor [a]	2.2	0.24
Single sensor [b]	3.2	0.34
Die to die [c]	5.2	0.56
One wafer [d]	9.6	1.0

(a) *No alignment between measurements, no saddle point value.*
(b) *Chip reloaded and new alignment, saddle point value.*
(c) *Saddle point value.*
(d) *Single measurement value (no saddle point value).*

cal alignment is accurate. This will result in a sensitivity deviation as the result of a placement error. This results in a sensitivity variation of 1 %. The variation is almost cut in half if a saddle point value is extracted. The die to die and one wafer characterization include all 50 selected dies. If the sensitivity characterization is restricted to only one die, the resulting standard deviation is smaller than 0.5 %.

The above sensitivity characterization is performed under the assumption of a constant force at the capillary tip. Thus, the measurements are performed with the same capillary without any capillary exchange during the measurements.

Capillary changes or transducer horn replacements will result in much higher variations, as the force at the capillary is no more constant for identical transducer current amplitudes. Therefore, ultrasound force calibrations with microsensors can improve the process stability as demonstrated in [4]. The transducer current is in this case adjusted so that the force at the tip remains constant during capillary or transducer horn changes.

Figure 4.9 shows the relative sensitivity of the sensors on the multiplexer in comparison with the single sensor located in the center of each die (see Fig. 3.1). The ultrasound caused force signal of the inner row is in good agreement with the signal strength of the single sensor whereas sensors of the outer row show significantly lower or higher sensitivity. Sensors that lie at the chip border parallel to the ultrasound direction have higher sensitivity and sensors at the upper or lower border have a lower sensitivity. This sensitivity change of sensors that lie close to die corners can be explained with changes in the stress field distribution due to the small distance to the corner.

The slightly higher sensitivity of sensors located at the inner row compared to the single calibration sensor is attributed to a different determination of the sensor voltage. The sensitivity change of sensors at the die border is systematical and is there-

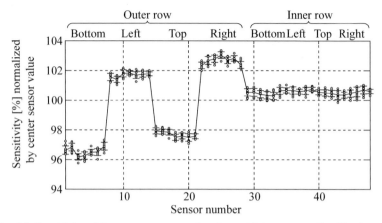

Fig. 4.9. Sensitivity of the sensors connected to the multiplexer normalized by the sensitivity of the single center sensor of the same geometry. For each sensor 10 measurements were performed.

fore rectifiable. This border sensitivity dependence is automatically adjusted by the measurement software.

4.2.6   Temperature Dependence of Sensitivity

As the to be examined packaging processes are covering a wide temperature range, the temperature dependence of the sensitivity and offset of the sensor is an essential sensor characteristic. Figure 4.10 shows the measured temperature dependence of the sensitivity as function of the substrate temperature. The temperature coefficient of the sensitivity (TCS) is extracted by a linear fit. The measurements are performed at ultrasound frequency. Bond force and ultrasound amplitude of the freely oscillat-

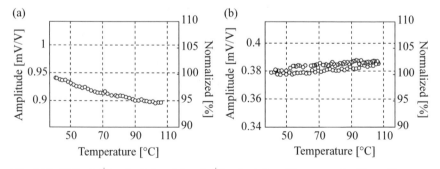

Fig. 4.10. TCS of p^+ implantation (**a**) and n^+ implantation (**b**) measured with capillary imprints on sensor $XYZ\text{-}Pass_{aligned}$ (**a**) and sensor $XYZ\text{-}Pass_{oblique}$ (**b**), respectively.

ing horn amplitude are 500 mN and 150 nm, respectively. The die is bonded with silver filled epoxy on the custom made ball grid array (BGA) substrate. The substrate temperature is measured with an on-chip integrated serpentine shaped aluminum resistor based on the MET1 metallization of the CMOS layers.

The temperature coefficient of sensitivity (TCS) of the x/y-force elements of sensor XYZ-$Pass_{aligned}$ with p$^+$ implantation resistors is -720 ± 100 ppm/K at 25 °C. The TCS of the x/y-force elements of sensor XYZ-$Pass_{oblique}$ with n$^+$ implantation resistors is 280 ± 100 ppm/K at 25 °C. If the chip temperature crosses about 140 °C nonlinear behavior in the sensitivity is observable. During the subsequent cool down cycle sensitivity fluctuations of several percent are observable. This behavior of the sensitivity correlates with offset hysteresis effects.

The TCS of the XYZ-$FC_{aligned}$ sensor determined with the shear test calibration is approximately -700 ppm/K and in accordance with the measurements at ultrasound frequency.

4.3   Linearity

Figure 4.11 shows the linearity between the amplitude of the ultrasound current of the transducer horn and the tangential force measured by the microsensor XYZ-$Pass_{aligned}$. Both amplitudes are extracted by the use of a quadrature filter at the fundamental frequency of the ultrasound oscillation (131.2 kHz). The ceramic imprint capillary TAMP-C-.001"-XL is used for the presented linearity correlation. The applied bond force is 600 mN. The ultrasound generator is operated in the constant voltage mode. A slow ramp up profile of the ultrasound amplitude is selected for the measurement. The horn amplitude is measured with a laser interferometer at the capillary shaft just underneath the horn. The amplitude measured with a freely oscillating capillary is 2.44 ± 0.16 μm/A. The slope in Fig. 4.11 is 15.6 ± 0.2 mV/V/A or 6.4 ± 0.4 mV/V/μm, correspondingly. Both values depend on the capillary

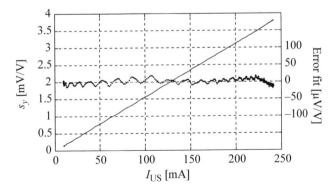

Fig. 4.11. Linearity between ultrasound current and sensor signal s_y for capillary imprints on passivated pads.

type and capillary fixation condition as e.g. fixation height. The remaining oscillations in the fit error are caused by ultrasound current signal fluctuations. Higher ultrasound amplitudes will initiate sliding between the capillary and passivation surface resulting in surface abrasion of both the capillary tip and the silicon nitride passivation.

4.4   Placement Sensitivity

The unique separation of the forces along the different axes is only valid for ideally centered circular contacts. As a matter of fact, the placement accuracy of the used wire bonding equipment is specified to ±3.5 μm. Moreover, if capillary imprints on passivated pads are employed, the exact position of the contact zone between a wire bonding capillary and the passivation is inherently unknown. Uniaxial force measurement for second bonds is no more guaranteed as the contact of a second bond does no more fulfill the symmetry conditions. All these non-idealities result in cross sensitivities between the different axes and placement dependence of the sensitivity. The strength of these effects has to be considered for correct sensor signal interpretation. The cross-sensitivity and placement dependence is described by the constants g_{ijk} and g_{ijkl} as introduced in the Sect. 2.1.

The wire bonder offers an elegant method to experimentally verify cross-sensitivity and the placement dependence. Figure 4.12 shows signal strength values of the x-, y-, and z-force sensor as a function of the position of the contact recorded on

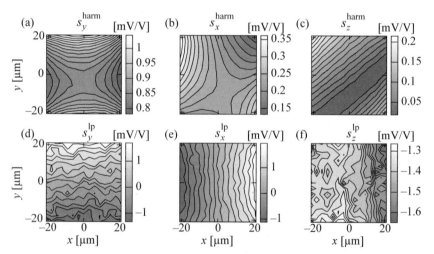

Fig. 4.12. Sensitivity surface of sensor XYZ-$Pass_{aligned}$ measured with the imprint capillary TAMP-C-.001"-XL. The upper three graphs are signal amplitudes (absolute value) at ultrasound frequency, the lower graphs are low-frequency signal values caused by the applied bond force. The used transducer horn has a relatively high x-force oscillation ratio of 27.7 %. The saddle point of the y-force signal is at (−1.5 μm, −2.9 μm) with a signal strength of 0.86 mV/V. The ultrasound force is applied in y-direction.

a passivated sensor with imprint capillary TAMP-C-.001"-XL. The sensor signal strength is measured at 225 different positions on a 15 × 15 array. The grid spacing of the scanned surface is 3 μm. More information on the recording process is found in Sect. 4.2.3. Bond force and ultrasound amplitude are 500 mN and 150 nm, respectively. The absolute alignment between sensor center and the wire bonder coordinate system is bound to the uncertainty in the knowledge of the exact contact position even if the optical alignment system of the wire bonder offers a high alignment accuracy. The upper three contour plots (a) to (c) of Fig. 4.12 show the sensor signal amplitude at ultrasound frequency. An ideal behavior of the ultrasound and sensor systems would result in vanishing x-forces and z-forces.

It is often found that the ultrasound oscillation of the capillary is not entirely in parallel to the y-axis. The fraction of the ultrasound amplitude in transverse direction strongly depends on the transducer horn. The ultrasound system of the used wire bonding machine has a relatively high transverse oscillation of about 25 %. The fraction of the ultrasound x-force is extracted at the saddle point position of the y-force signal. Another reason for non-uniaxial ultrasound sensing on chip is die rotation as a result of the die bonding process.

The sensor signal of the x-force sensor is thus a superposition of cross-talk of the strong ultrasound force in y-direction and the x-force due to the ultrasound oscillation in transverse direction. Simulated values for cross-talk and x-force are given in Fig. 4.13. This simulation is based on the parametric electrical simulation flow explained in Fig. 2.16. The varied parameter is the contact position. The strength of the x- to y-force ratio is extracted from measurements. The stress field used for the simulation is based on FEM data of a layered anisotropic half space as described in Sect. 2.1.3. The piezoresistive coefficients of the simulation are $\pi_{11} = 1.1 \cdot 10^{-11} \, \text{Pa}^{-1}$, $\pi_{12} = 1.0 \cdot 10^{-11} \, \text{Pa}^{-1}$, and $\pi_{44} = 53 \cdot 10^{-11} \, \text{Pa}^{-1}$.

The superposition agrees well with the measured signal amplitude of the x-force sensor. As the ultrasound oscillation in z-direction is small, the z-force sensor signal is dominated by x- and y-force cross-talks.

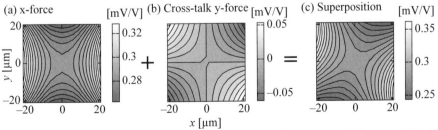

Fig. 4.13. Simulation of the x-force component (**a**) of the x-force sensor and the cross-talk of the y-force (**b**). The weighted combination of both surfaces is in good agreement with the x-sensor sensitivity surface as measured in Fig. 4.12. The tangential ultrasound y-force amplitude is set to 82 mN. The x-force is set to 27.7 % of the y-force amplitude.

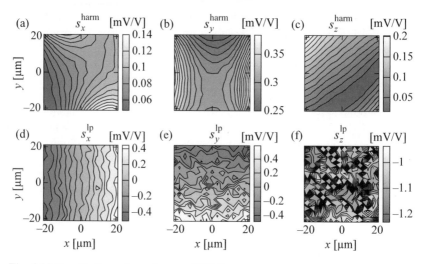

Fig. 4.14. Sensitivity surface of sensor $XYZ\text{-}Pass_{oblique}$.

The lower three graphs (d) to (f) of Fig. 4.12 are sensor signal contour plots of the low-frequency signal raise caused by an applied bond force. An ideal behavior of the wire bonding system and the sensor system would solely result in an offset for the z-force sensor that is constant over the different bond positions and proportional to the applied bond force. The sensitivity surface of the x- and y-force sensor is primarily reflecting the cross-talk of the applied bond force. The bending of the transducer horn under an applied normal force results in a force component in y-direction. Due to sliding effects during the impact phase, most of this y-force is translated into a y-displacement. The fluctuations in the low-frequency y-force signal are the residuals of this coupling. As a consequence, measured low-frequency x- and y-forces have to be carefully interpreted due to the cross-coupling of the applied normal force. Under the condition of a dominant ultrasonic y-force the contact position can be extracted from a saddle point calculation of a pure quadratic fit of the y-force signal.

Similar placement dependence surfaces and cross-sensitivity properties are obtained for other sensor designs as shown in Fig. 4.14 for sensor $XYZ\text{-}Pass_{oblique}$.

4.5   Offset

Offsets and especially the temperature coefficient of the offset need consideration for the flip-chip application. To estimate the error due to temperature induced offset drifts, measurements were performed on unflipped dies. The signal of a sensor surrounding an unloaded ball is ideally zero. Several effects result in an unbalanced Wheatstone bridge and thus in an offset of the sensor. These effects are often dependent on temperature.

An obvious cause of sensor offsets is unsymmetrical wiring of the sensor elements. The effect of the wiring resistances is examined for sensors S_{outer}^{top}. The orientation of the sensor connecting structures is shown in Fig. 4.15. The outer sensor connecting structures are identical and are rotated by a multiple of $^\pi/_2$. The sensing lines are connected to the multiplexer so that all y-force signals are routed to the same output line. The inner sensor connecting structures slightly differ from the outer connecting structures and are mirrored.

The sensor design for the wire bonding application is space limited. It is desired to achieve a high test pads density on a small silicon die. The wiring of the sensors takes up a relatively large fraction of the whole sensor extension as only two metal layers are available.

To get a compact sensor design, the x-, y- and z-force sensors are powered with common supply distribution lines. The use of via contacts is minimized and equally distributed on the different power and sensing lines. Currents of up to 6 mA in combination with the resistance of the metal tracks causes significant voltage drops along the power lines. Figure 4.16 shows the sensor design of sensor $XYZ\text{-}Pass_{aligned}$ with a superposed equivalent resistor network of the power lines. The resulting sensor offset of the x-force, y-force and z-force sensor are 0.76 mV/V, 0.79 mV/V, and 16 μV/V, respectively. The offsets of the sensors situated at the inner ring are of comparable magnitude.

If space is of minor concern, the sensor in the different axis direction should be powered with individual supply lines connected to a hub. Different distances between sensor lines can be compensated for by serpentine lines. This balancing is especially important for further reduction of thermal offset drifts.

(a)

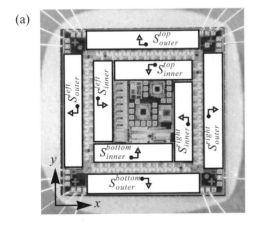

(b)

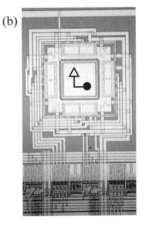

Fig. 4.15. Orientation of the sensors (**a**) on the test chip (**b**). The sensor signal lines are connected to the multiplexer bus so that sensors that are receptive to forces in the same directions are connected to the same output lines.

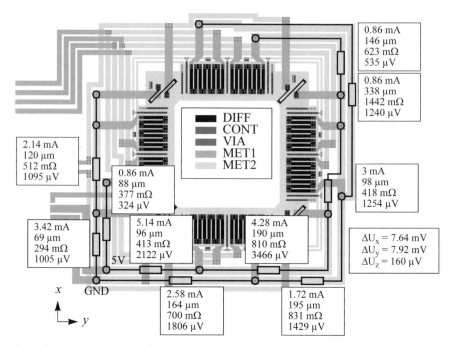

Fig. 4.16. Estimated voltage drops along power supply lines for the sensor *XYZ-Pass*_{aligned}, resulting in an offset of the sensor voltage.

In reality, much larger offsets are found for the different sensors. Most striking are offsets on the z-force sensors that can exceed 7 mV/V. In addition, a large die to die variation of the offset is observed. To further examine offset sources, the offset distribution on the wafer was recorded. About 50 dies are picked from different positions of diced wafers and glued on the custom made BGA with Loctite 480 adhesive. All offset and offset drifts of the sensors on the die are measured. The leadframe indexing system of the wire bonder in combination with the custom made downholder allows these measurements to be performed with reasonable time effort. A special low noise, low drift, low offset preamplifier stage was used for this offset characterization.

Figure 4.17 shows the offset distribution of the z-force sensors S_{outer}^{top} and S_{outer}^{left} as function of the die position on the wafer. The length of the arrows indicates the offset value of the sensor located at the start point of the arrow. The sensor groups (S_{outer}^{top}, S_{outer}^{left}, ...) have identical offset distributions for all top/bottom and left/right positions, respectively. The offset of the sensor is composed of a constant offset value and a spatially varying part. The mean offset value of the z-force sensor is –2.75 mV/V at 36 °C. This constant part of the z-force offset has a large temperature coefficient of –10 ± 0.2 µV/V/K. There is no indication of an influence of the

(a)   (b)

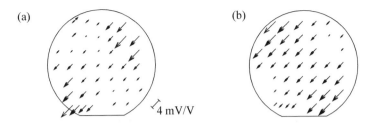

4 mV/V

Fig. 4.17. Offset distribution of the z-force sensor on a wafer. Vector fields (**a**) and (**b**) show the offset of the top/bottom sensors S_{outer}^{top}, S_{outer}^{bottom}, S_{inner}^{top}, S_{inner}^{bottom} and the left/right sensors S_{outer}^{left}, S_{outer}^{right}, S_{inner}^{left}, S_{inner}^{right} respectively. The arrow orientation is arbitrary. For orientation of the sensor structures see Fig. 4.15.

die edge on the z-sensor offset as inner and outer sensors have identical offset characteristics. Possible explanations of the gradients in the offset across the wafer are stress field gradients or processing related variations. Stress field caused offset gradients can be ruled out by the different behavior of sensor groups that are rotated by $\pi/2$.

Similar offset gradients on the wafer are found for the x- and y-force sensors as shown by the vector field in Fig. 4.20a. Each vector is the x-/y-force offset averaged over all sensor of the die at the position of the arrow. The offset difference of sensors positioned at opposite sides of the die is mainly caused by the unbalanced design.

4.5.1 Temperature Coefficient of Offset

Thermal drift properties of the sensor are of notable importance for the flip-chip application. Especially, the recording of forces during thermal cycling demands an understanding of the various drift effects.

The temperature dependence of the sensor sensitivity is addressed in Sect. 4.2. In contrast, this section elaborates on the thermally induced drift behavior of the sensor offset. The temperature dependence is characterized in first order with a temperature coefficient of the offset (TCO) for predictable effects. Overstressing of the device can result in random signal fluctuations that can only be characterized with error bounds.

First measurements showed, that the die fixation heavily affects thermal behavior of the offset. Figure 4.18 compares the TCO for different die mount techniques. The configuration (a) is used for the wire bond application. The die is bonded on the BGA substrate with silver filled epoxy adhesive. Epoxy residues form a fillet around the die. Offset measurements of dies with an epoxy fillet entail large TCOs for sensors close to the die edge. Typical values of the outer sensors are 16 µV/V/K. The x- and y-component of the vectors in Fig. 4.18 indicate the TCO of the x- and y-force sensor, respectively.

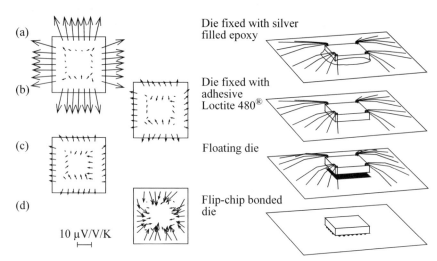

Fig. 4.18. Offset TCs for different configurations. Die bonded with epoxy adhesive (**a**), die bonded with small amount of Loctite 480® adhesive (**b**), released die (**c**), and flip-chip bonded die (**d**).

The temperature dependence of the offset can be significantly reduced by using a smaller amount of adhesive so that only the backside of the die is covered. Again the adhesive Loctite 480 of Henkel Loctite, Munich, is used for the offset characterization. Further improvement of the offset is achieved for completely released dies. The following procedure was used to prepare such floating dies (see Fig. 4.19): the chip is mounted on a wafer dicing tape and then fixed on the chip carrier or leadframe with a double sided tape. For substrates that are difficult to bond as e.g. flip-

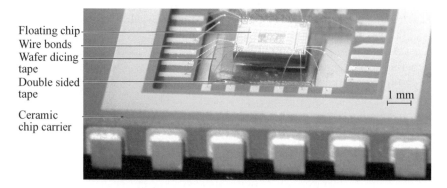

Fig. 4.19. Floating chip as used for offset measurements. For wire bonding the chip is fixed on a wafer dicing tape that is turned over and fixed on the chip carrier with a double sided tape.

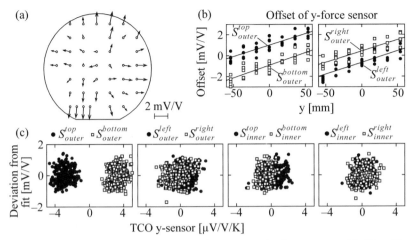

Fig. 4.20. Sensor offset distribution on wafer for x- and y-force sensors (**a**). The offset voltage difference of sensors lying on opposite edges of the die is constant over the wafer (**b**). The TCO does not depend on the offset gradient on the wafer (**c**). The TCO value is a function of the sensor position on the die.

chip dies with a gold flash on the Ni layer, the double sided tape had to be substituted by a glue to increase the stability for high bonding temperatures. After wire bonding, the die is released from the wafer dicing tape. The die is henceforth only supported by the wire bonds.

The offset of the y-force sensor as function of the die position on the wafer is shown in Fig. 4.20a (y-component of the arrows). The offset values of the different sensor groups can be described by parallel lines if plotted as function of the y-position (see Fig. 4.20b). A temperature change will shift the lines to higher and lower values. The y-force sensor offset gradient in y-direction remains unchanged. The TCO of floating dies is therefore uniform over the different sensor groups as defined in Fig. 4.15. The TCO vector is mainly directed perpendicular to the closest die edge. This behavior is consistent over the wafer as shown in Fig. 4.20c. There is no correlation between the offset distribution on the wafer and the corresponding TCO. The measurements shown in Fig. 4.20 are performed with wire bonding test chips with Al-pads. Measurements on flip-chips with solder balls show equivalent TCOs as shown in Fig. 4.21. Figure 4.21 shows the TCO distribution across the pads of the die. Obviously, the TCO value is a function of the sensor position on the die. It is possible to correct a part of this TCO. The remaining variation of the TCO is 3 µV/V/K or 0.3 mN/K, correspondingly.

Figure 4.18d shows the TCO for a flip-chip bonded die, which is caused by the mismatch of the thermal expansion coefficients (TCE) of the substrate and the die. A comparison between the flip-chip force signals of bonded dies (see Fig. 4.18) and the TCO points up the importance of exact TCO knowledge to get an accurate measurement of the forces during cycling.

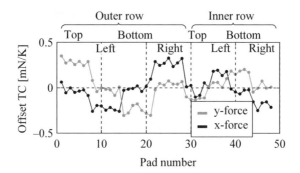

Fig. 4.21. Temperature coefficient of the force-equivalent offset (TCO) as function of the pad number for a flip-chip test die with solder balls on the test pads. The flip-chip is not bonded to a substrate.

Previously mentioned offset variations are predictable if the TC of the offset is known. Moreover, these temperature originated offset variations are reversible. In addition to this reproducible offset behavior, fluctuations in the sensor signal can be observed for large temperature differences. Figure 4.22 shows the offset of sensors on floating dies during temperature cycling. The measurement is recorded in a temperature chamber VT7010 of Vötsch Industrietechnik GmbH, Balingen, Germany. The temperature is measured with PT100 resistors. The dashed line is the temperature measured by the PT100 resistor mounted above the floating die. The black and gray lines are the y-force sensor signals of sensor 3 and 7, respectively. Cycling with a temperature range of 58 K results in a reversible offset behavior. An offset drift is observed for temperature cycles larger than about 80 K, resulting in a non-reversible offset behavior. A possible explanation of this fluctuations is the plastic behavior of the aluminum layer that covers the sensor structure. The elastic and plastic behavior of aluminum during thermal cycling is described in [8] for silicon wafers covered with aluminum and a SiO_2 barrier layer.

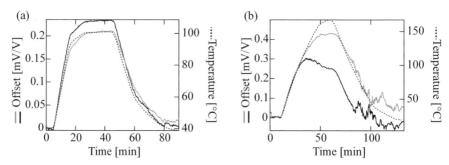

Fig. 4.22. Offset drift of the y-force signal of sensor 3 and sensor 7 for thermal cycling between 22 °C and 80 °C (**a**), 167 °C (**b**), respectively.

The strength of the hysteresis and the fluctuations of the x-/y-force sensors depends on the thermal cycling range as shown in Fig. 4.23. Such a behavior was especially pronounced for pads covered with a thick Ni-Au metal stack. This offset hysteresis can not be compensated for during a measurement and therefore limits the accuracy of the force measurements. Sensors near the die corner will further-more sense temperature dependent stress gradients due to bi-metal stresses arising from mismatches in the thermal expansion coefficients of the CMOS layers [9]. The exponential increase of this effect towards the die edge will become critical if the sensors are placed at close distance to the die edge.

The aspired temperature range for wire bonding real-time measurements is 20 °C up to 180 °C and for the flip-chip applications –40 °C to 125 °C. Although highest possible insensitivity to temperature and intrinsic stress fields is achieved with the use of a full Wheatstone bridge configuration, temperature effects are still the limit-ing factor in the sensor performance. Plastic flow in the metallization layers of the chips result in force hysteresis curves during thermal cycling above 90 °C.

For the wire bonding application, the signals at ultrasound frequency are of main interest. Offset changes are therefore negligible as long as the sensor offset is still in the measurement window of the signal recording equipment. Thus, the sensors can be used for temperatures up to 180 °C.

Tables 4.6 and 4.7 summarize offset origins and the temperature coefficient of the offset (TCO) for the x/y-force and z-force sensors, respectively. Values that are characterized with a "lesser than" symbol indicate upper values for the property (absolute value). More accurate determination of the property would need a larger sample count. The sensor to sensor variation of the TCO in combination with the random signal fluctuations gives an estimation of the possible error during thermal cycling measurements.

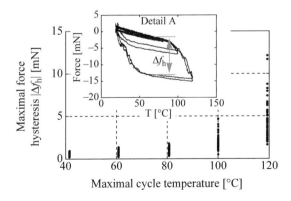

Fig. 4.23. Maximal force hysteresis $|\Delta f_h|$ of unflipped dies for various cycling temperature ranges. Detail A shows the extraction of Δf_h from one cycling curve.

Table 4.6. Offset sources of the x- and y-force sensor.

Offset sources x- and y-force sensor	Inner row		Outer row	
	Offset [mV/V]	TCO [μV/V/K]	Offset [mV/V]	TCO [μV/V/K]
Wafer processing	−1.5 … 1.5	< 0.5	−1.5 … 1.5	< 0.5
Sensor position dependence on die	< 1	−2 … 2	< 1	−4 … 4
Sensor to sensor variation	< 1	< 1	< 1	< 1
CMOS layers	< 1	< 1	< 1	< 1
Die bonding epoxy adhesive	(a)	< 1.5	(a)	−16 … 16
Design, longitudinal to edge [b]	0.706	(a)	0.765	(a)
Design, transverse to die edge [b]	0.876	(a)	0.792	(a)

(a) Value not determined.
(b) Theoretical value.

Table 4.7. Offset sources of the z-force sensor.

Offset sources z-force sensor	Inner row		Outer row	
	Offset [mV/V]	TCO [μV/V/K]	Offset [mV/V]	TCO [μV/V/K]
Wafer processing	−4 … 4	< 1	−4 … 4	< 1
Sensor position on die	(a)	−1 … 1	(a)	−1 … 1
Sensor to sensor variation	< 1	< 1	< 1	< 1
CMOS layers (*XYZ-Pass*)	−2.75	−10	−2.75	−10
(*XYZ-Al85*)	(a)	−2.5	(a)	−2
(*XYZ-FC*)	(a)	—	(a)	1
(*XYZ-Au85*)	(a)	0.5	(a)	0.5
Die bonding epoxy adhesive	(a)	< 1	(a)	< 1
Design [b]	0.016	(a)	0.016	(a)

(a) Value not determined.
(b) Theoretical value.

The following conclusions can be made on the basis of the observed offset characteristics:

- Although the offset values of the x- and y-force sensor depend on the die position on the wafer, no TCO position dependence could be measured.
- The TCO of the x- and y-force is independent of the pad material and structuring.
- Under absence of external influences as e.g. epoxy fillets, the TCO values of the x- and y-force sensors depend mainly on the sensor position on the die. This value can be compensated for, as it is the case for flip-chip measurements.
- The TCO of the z-force sensor depends heavily on the pad structure. In particular, the TCO of passivated pad structures is exceptionally large.
- The offset of z-force sensors depends strongly on the position of the die on the wafer. A constant offset value is superimposed on the offset gradient on the wafer. This constant offset value entails a large TCO.

4.6   Summary of Technical Data of the Test Chip

In the following the characteristics of the developed sensor $XYZ_{aligned}$ are summarized.

Process		0.8 µm double metal, polysilicon CMOS
Power consumption		6 mA @ 5 V sensor voltage
Sensitivity	x-/y-force	10.2 ± 0.5 mV/V/N
	z-force	2.24 ± 0.05 mV/V/N
TCS	x-/y-force	−720 ppm/K
	z-force	280 ppm/K
TCO	x-/y-force	< 5 µV/V/K
		< 3 µV/V/K (compensated)
	z-force	< 2 µV/V/K (FC pads with solder balls)
NEF	x-/y-force	380 nN$/\sqrt{Hz}$ @ 5 V sensor voltage
Placement suscept.	x-/y-force	±8.2 µm to achieve 2 % accuracy
Cross-talk	x ↔ y	$g_{xyxy} = 1.5$ µV/V/N/µm^2
	z → x, z → y	$g_{xzx} = 127$ µV/V/N/µm

Test pads

Active test pad number	48
Metallizations	Al (1.5 µm)
	Al/Ni/Au (1.5 µm / 5 µm / 1 µm)
	Al/Ni/Au (1.5 µm / 5 µm / 50 nm)
	Si_3N_4
Maximal pad size	85 × 85 µm^2

Dimensions

Chip	3 × 3 mm^2

5 Applications

The microsensors presented in the earlier chapters have initially been developed to investigate wire bonding processes. The measurement system presented in Chap. 3 offers the possibility of recording a full time-dependent signal set of the x-, y- and z-forces in combination with the wire bonder signals z-axis position, ultrasound current, and bond force. These signals provide information about the bonding process in conjunction with machine characteristics. The high repeatability of the in situ and real-time measurements also adds to the wire bonding machine analysis under real process conditions. Distinct applications have been defined and are used to support the development of new wire bonders. They are described in Sects. 5.1 to 5.3. The success of the investigations suggested an extension of the measurement method to flip-chip bond contacts, as described in Sect. 5.4. An overview of all applications is given in Fig. 5.1.

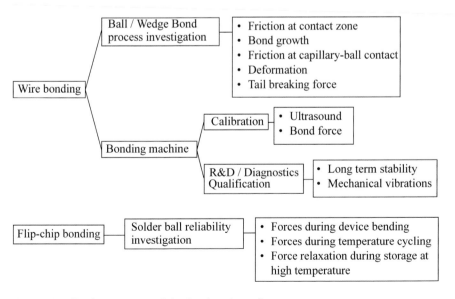

Fig. 5.1. Application spectrum of the developed xyz-force sensors.

5.1   Wire Bonder Development

Following the trend of miniaturization, wire bonder manufacturers like ESEC, Cham, Switzerland, are spending much effort on their bonders' fine pitch capability. In the late 1990s, a bond pad pitch of 80 μm was the standard in mass production wire bonding, while in the first few years of the new millennium, the state-of-the-art pitch is 50 μm, with 40 μm being ramped up in a larger scale. The challenges of ever more miniaturized bonding processes are the need for higher positioning accuracy accompanied by the smaller bond parameter windows, where parasitic effects can become dominant. Bonder to bonder variations acceptable for earlier processes become too large for the miniaturized processes. Among the ball bonder modules continuously improved to allow for ever more accurate process parameters, are the bond head positioning system, the flame-off system, the ultrasonic system, and the bond force system.

Each newly developed module goes through a functional test, separated from the bonder, followed by the integration test on the bonder. While the functional test involves electronics testing and mechanical measurements, the integration test covers the main wire bonder function, the bonding process.

The best process test is to bond similar products like the prospective customers and check if a specific bond quality is reached. For the initial bond quality, the standard tests are shear, pull, and geometry testing. The final bond quality is determined by reliability tests such as harsh environment and temperature cycling tests of the encapsulated devices. For the average bonder manufacturer it is impractical to perform the same effort for tests and reach the same specifications like the wire bonder end user, as test equipment and methodologies, bond material, and cleanroom classes are differing considerably. Prior to releasing a new module, its long term stability needs to be assessed. For this, a complete bond production process has to be run during days or longer, with the bond quality continuously monitored. Moreover, the chosen test material should be representative for a majority of products. Such a requirement can be impossible to fulfill because sometimes the industry shifts to completely new applications during the development of a new bonder. In summary, full validity of the process test can be too difficult to achieve for bonder manufacturers, and the microsensor technology described in this book presents a viable alternative.

From a mechanical point of view, the xyz-force sensor offers measurements of highly process relevant quantities: forces between capillary tip and chip which are acting on the wire to deform it and on the interface to produce the bond. Obviously, metallization and wire material as well as temperature and contaminations can have even larger influences on the bonding result, but depend on the specific bonding application. Force variations and parasitic vibrations at the bonding place, however, can impair all applications. If measured by microsensors, these signals allow for fast monitoring and testing, and are a substitute for extended process tests. No other measurement technology allows for a comparable combination of real-time and in situ signals to be recorded. For ESEC, the microsensor technology has continuously been a valuable tool for speeding-up the development of new bonders, and three of the used applications are outlined in the following subsections: module calibration, vibration (speed) characterization, and long term stability testing.

5.1.1   Process Module Calibration

When producing wire bonds in large numbers, more than one wire bonder is usually needed to obtain a high throughput of units. As the optimization of the bonding parameters is a time consuming task, it is usually done on one wire bonder only. Then it is desired to use the optimized parameters also for the other wire bonders. This is possible only if the wire bonders are similar to each other, i.e. if especially the process modules for ultrasound and bond force are calibrated.

The main process responses are bonded ball diameter (BDC), height, and shear strength. The yield of a production process depends on the chosen upper and lower limits of process responses, and on the distributions of the process responses. These variations of process responses are caused by bond material variations and variations of the input parameters. Here, the focus is on the reduction of the latter.

The basic input parameters of the thermosonic ball bonding process are bond force, ultrasound value and time, and chip temperature. It is the goal to build wire bonders to control these parameters as accurately as possible to reduce the need of calibrations. For advanced fine pitch processes, however, the calibration of the ultrasound and bond force modules remains an option to increase the yield.

Ultrasound Calibration. The quantities ultrasound current I_{US}, vibration amplitude of horn A_H, and vibration amplitude of capillary tip A_C, as defined in Sect. 1.2, are directly proportional to each other with proportionality constants $p_1 = A_H/I_{US}$, and $p_2 = A_C/A_H$. In the absence of friction at the interface, i.e. for high bond forces and low ultrasound parameters, and for a given I_{US}, F_T is directly proportional to the amplitude A_C measured earlier in free air, and $p_3 = F_T/A_C$.

When transferring a set of process parameters from the optimized bonder to the next, variations of the ultrasonic system arise from the horn (p_1) and capillary (p_2, p_3) amplifications. The process is affected by a new set of proportionality constants, differing by Δp_1, Δp_2, and Δp_3, respectively. There are various calibration methods to control such ultrasound parameter variations. Among them are optical measurements using either a laser interferometer (laser vibrometer) or a light emitting diode (LED) with a position sensitive detector (PSD) to measure the ultrasonic amplitude of the horn tip, A_H, as shown in Fig. 5.2a. This calibration is used on production

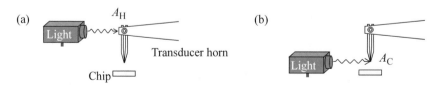

Fig. 5.2. Optical measurement of ultrasonic amplitude at horn tip (**a**) and capillary tip (**b**).

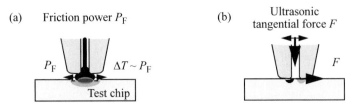

Fig. 5.3. Calibration methods with integrated (**a**) temperature and (**b**) force sensors.

floors [1]. It accounts for the variations of the horn amplification factor, Δp_1. A further improvement of the calibration is expected if measuring the capillary tip amplitude, as shown in Fig. 5.2b, because in addition to Δp_1, the capillary amplification variations Δp_2 will be cancelled, too [2].

It was found that sub 50 μm fine pitch process responses strongly depend on Δp_3 [3] and other capillary influences such as the variations of the inner hole and chamfer diameters, and the face angle. The issue can be tackled by introducing tighter capillary tolerances, resulting in higher capillary cost, or by introducing a new calibration method to account for Δp_3. Two such methods based on microsensors are presented in the following. The methods are based on measurement of (a) the friction power [4] and (b) the tangential ultrasonic force [5, 6], as illustrated in Figs. 5.3a and (b), respectively.

The first approach uses a resistive temperature detector, designed as shown in Fig. 5.4a, to measure the temperature increase proportional to the ultrasonic friction power generated by a ball bond on a passivated sensor surface. An SEM micrograph of the whole sensor structure is shown in Fig. 5.4b. The friction power, a function of

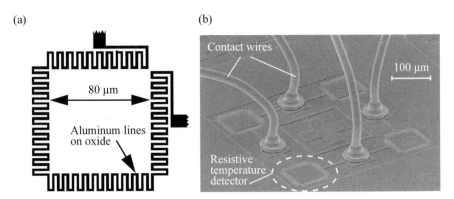

Fig. 5.4. Schematic layout (**a**) of the temperature microsensors: an integrated aluminum resistor is used as resistive temperature detector. Microsensor (**b**) as a part of a Wheatstone bridge with four identical resistors. Wire bonds connect to the bridge circuit.

the resistance change of the Al-resistor, depends on the ultrasound parameters, namely the free vibration amplitude of the capillary, as shown in Fig. 5.5a. Unfortunately, variations in bond force will also change the friction power in a way that is difficult to account for, as shown in Fig. 5.5b. A further drawback of the temperature sensor method is that it needs a wire. The balls have to be placed somewhere after the measurement. Such a method is impractical in a production environment.

An alternative idea worked out at ESEC is to cancel capillary stiffness variations by calibrating the ultrasonic tangential force using an integrated xyz-force sensor as shown in Fig. 5.6a. The sensor is similar to that described in Sect. 4.2.3. The test zone of the sensor is covered by the whole dielectric layer stack of the CMOS process. The y-force calibration factor was determined using the shear tester method with a ball bond on a metallized test zone at ambient temperature, as described in Sect. 4.2.1. The resulting experimental calibration value is 10.4 mV/V/N.

The sensitivity of the tangential force sensor depends considerably on how exactly the capillary tip is centred on the test zone and how ideally the contact ring transmits the tangential force to the chip. An example imprint of a capillary on an Al-pad is shown in Fig. 5.6b. To account for placement and contact ring variations, several measurements are carried out, each with the capillary aligned on one point of a predefined grid, as described in Sect. 4.2.3. Examples of a y-force sensor signal and the signal amplitudes obtained at each grid point are shown in Figs. 5.7a and (b). The ultrasound tangential y-force value F is taken from the saddle point of the response surface. This value is used to carry out the ultrasound calibration. The measured F error of repeatability is below 2 %. The measured x-force F_x typically is about 1.5 % of F for the used horns and neglected in the following. A standard capillary is used. The degradation of the test zone surface or the capillary tip has been investigated and found to be negligible for the number of measurements needed.

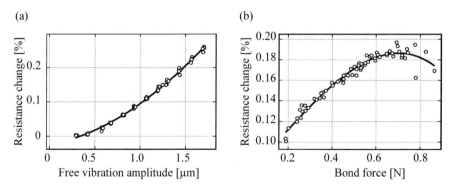

Fig. 5.5. Temperature induced resistance change due to ultrasonic friction power versus ultrasonic vibration amplitude (**a**) and bond force (**b**).

(a)

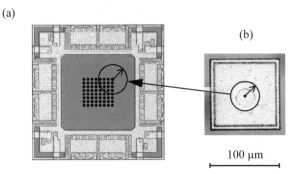

(b)

100 µm

Fig. 5.6. Grid for capillary touchdowns for calibration procedure (**a**). Capillary imprint on pad (**b**).

To characterize the potential of the calibration method, a test bond process on conventional test chips was used. A bottle-neck capillary type suitable for 50 µm ball pitch in-line bonding and a 20 µm diameter AW-88 wire were selected. The chip temperature was 160 °C. The average free-air ball diameter was 31.4 µm. Ball impact force, bond force, and ultrasound time were 130 mN, 100 mN, and 10 ms, respectively. Wedge impact force, bond force, and ultrasound time were 400 mN, 300 mN, and 9.3 ms, respectively.

The process was designed to show a strong sensitivity to ultrasound variations. On each chip, nine balls are bonded with ultrasound currents varying between 150 mA and 450 mA inducing large variations of the ball geometry. Examples of such deformed balls are shown in Fig. 5.8. The BDC was chosen as an indicator for the ultrasound effect. To cancel out chip-to-chip variations, the BDCs from more

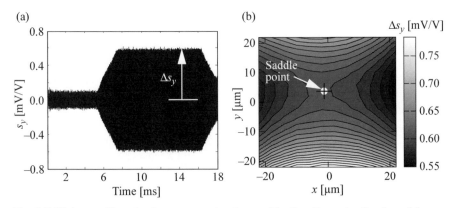

Fig. 5.7. High-pass filtered y-force sensor signal caused by the ultrasonic vibration of the capillary (**a**). Response surface showing y-force sensor signal as a function of the contact location (**b**).

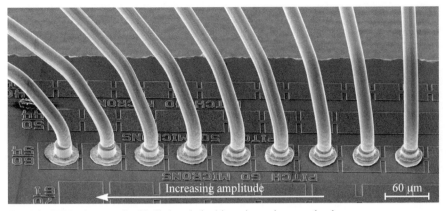

Fig. 5.8. SEM micrograph of balls bonded with various ultrasound values.

than 12 chips were averaged. A resulting deformation curve is shown in Fig. 5.9. To get a process measure for the ultrasonic force, the current target value, I^T, needed for a target BDC value, here defined as 40 μm, is evaluated.

A set of ten capillaries and four horns were used on one wire bonder. Capillaries and horns are selected that show higher ultrasound variations than usually accepted for productions in order to obtain a more pronounced effect of the calibrations. For each capillary/horn-combination $F_{h,j}$ is measured by the microsensor, where $h = 1...4$ and $j = 1...10$. The test process then is bonded on the test chips, and the target $I^T_{h,j}$ is determined. The measured values for $I^T_{h,j}$ are shown in Fig. 5.10a and plotted for each $F_{h,j}$ in Fig. 5.10b. The averages are denoted $\overline{I^T}$ and $\overline{F}$, respectively.

Five ways to run a production are compared in the following and denoted as procedures A, B, C, D, and E. Common to all procedures is the prediction of ultrasound currents and the assessment of the procedure (calibration) quality by comparing the

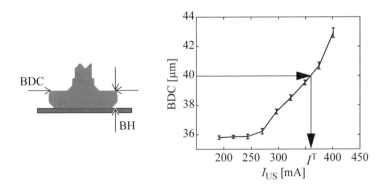

Fig. 5.9. Ultrasonic ball deformation.

predicted currents with the target currents. For this comparison, the standard deviation of the predicted from the target currents is used and named as "quality" of the procedure. The lower the standard deviation, the better the quality of the procedure. In contrast to procedures A and B which serve as references, the procedures C, D, and E each use the microsensor force values for calibration in a particular way. In the following, the five procedures are described in more detail, starting with the least effective, A, and ending with the most effective, E.

Procedure A solely consists of optimizing the ultrasound value with one capillary on one horn and then using this value for all other capillary/horn-combinations. As an example, the target current $I^{\mathrm{T}}_{m,n}$ for horn $m = 1$ and capillary $n = 1$ is chosen as predicted current, $I^{\mathrm{P}}_{1,1} = I^{\mathrm{T}}_{1,1}$, for all capillary/horn-combinations, i.e. $I^{\mathrm{P}}_{h,j} = I^{\mathrm{P}}_{1,1}$ for all h and j. The procedure quality is calculated as

$$\sigma_{m,n} = \sqrt{\sum_{h=1}^{4} \sum_{j=1}^{10} (I^{\mathrm{P}}_{m,n} - I^{\mathrm{T}}_{h,j})^2 / 40} . \tag{5.1}$$

Similarly, the target current of any other ($m \neq 1$ and $n \neq 1$) capillary/horn-combination, $I^{\mathrm{T}}_{m,n}$, can be chosen for the calibration, resulting in another quality value. Consequently, the average quality of procedure A is

$$\sigma_{\mathrm{A}} = \sum_{m=1}^{4} \sum_{n=1}^{10} \sigma_{m,n} / 40 . \tag{5.2}$$

For procedure B, the average of all 40 target currents,

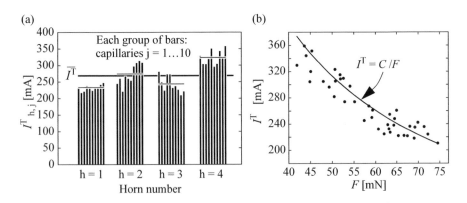

(a)   (b)

Fig. 5.10. Experimentally determined target currents to achieve a 40 μm BDC (**a**). Average values are indicated by horizontal lines. Transducer current (**b**) needed for BDC = 40 μm versus measured tangential force. Bullets are measurements for various horn-capillary combinations. Line is for $C = \overline{I^{\mathrm{T}}} \cdot \overline{F}$.

$$\overline{I^{\mathrm{T}}} \;=\; \sum_{h=1}^{4} \sum_{j=1}^{10} I_{h,j}^{\mathrm{T}} / 40 \;. \tag{5.3}$$

is used, resulting in the best possible single predicted current for all capillary/horn-combinations. The quality of procedure B is calculated as

$$\sigma_{\mathrm{B}} \;=\; \sqrt{\sum_{h=1}^{4} \sum_{j=1}^{10} (\overline{I^{\mathrm{T}}} - I_{h,j}^{\mathrm{T}})^{2} / 40} \;. \tag{5.4}$$

The two following procedures, C and D, use one predicted current for each horn which corresponds to a horn calibration. For procedure C, the microsensor measurement of the first capillary on the first horn, $F_{1,1}$, together with its target current is used to calculate a calibration factor, $C_{1,1} = F_{1,1} I_{1,1}$. Subsequently, for each horn, the first capillary is used to predict a current,

$$I_{h,1}^{\mathrm{P}} \;=\; C_{1,1} / F_{h,1} \;. \tag{5.5}$$

The quality of this procedure is calculated as

$$\sigma_{k} \;=\; \sqrt{\sum_{h=1}^{4} \sum_{j=1}^{10} (I_{h,k}^{\mathrm{P}} - I_{h,j}^{\mathrm{T}})^{2} / 40} \;, \tag{5.6}$$

where $k = 1$ denotes the first capillary. As capillaries 2 - 10 can also be used to predict the target current, the average quality of the procedure with $C_{h,j}$, where $h = 1$ and $j = 1$, is

$$\sigma_{h,j} \;=\; \sum_{k=1}^{10} \sigma_{k} / 10 \;. \tag{5.7}$$

As 40 different calibration factors $C_{h,j}$ are possible, the average of $\sigma_{h,j}$ is used as the quality value of procedure C

$$\sigma_{\mathrm{C}} \;=\; \sum_{h=1}^{4} \sum_{j=1}^{10} \sigma_{h,j} / 40 \;. \tag{5.8}$$

For procedure D, each average

$$I_h^P = \sum_{j=1}^{10} I_{h,j}^P / 10 \tag{5.9}$$

over the ten predicted currents for a horn is chosen as the predicted current for that horn. This corresponds to the best possible method to account for variations of p_1 without any ultrasound adjustments between capillary changes. The procedure quality averaged over all horns then is

$$\sigma_D = \sqrt{\sum_{h=1}^{4} \sum_{j=1}^{10} (I_h^P - I_{h,j}^T)^2 / 40} . \tag{5.10}$$

For procedure E, one calibration constant for each horn is determined using $C_h = \overline{F}_h \cdot \overline{I}_h$. For each capillary, a current is predicted using $I_{h,j}^P = C_h / F_{h,j}$, where $h = 1 \ldots 4$ and $j = 1 \ldots 10$. The standard deviations of the predicted from the target currents are obtained for each horn using

$$\sigma_h = \sqrt{\sum_{j=1}^{10} \frac{(I_{h,j}^P - I_{h,j}^T)^2}{10}} , \tag{5.11}$$

where $h = 1 \ldots 4$. The experimental results are shown in line 6, columns 2 - 5, of Table 5.1. The quality of procedure E is given as

$$\sigma_E = \sum_{h=1}^{4} \sigma_h / 4 . \tag{5.12}$$

Table 5.1. Standard deviations of chosen ultrasound currents from target [mA].

Horn Procedure	H 1	H 2	H 3	H 4	all	Yield Y [%]
A	43	50	47	65	57	78.92
B	36	31	35	59	42	86.23
C	19	41	38	34	33	91.74
D	9	30	24	21	21	98.54
E	14	13	9	19	14	99.95

The calibration after each capillary change according to procedure E accounts for variations of p_1, p_2, and p_3, reducing the standard deviations by an average of 33 % compared to that of the next best procedure, D.

To demonstrate the potential for process portability enhancement of the procedures, a 50 µm ball pitch process was used to translate $\sigma_{A\text{-}E}$ into yield values. Ball impact force and bond force were 120 mN and 80 mN, respectively. All the other parameters were the same as for the test process. BDC, ball height (BH), and shear strength (SS) are measured and used to calculate the Cpk-values for each ultrasound setting according to

$$\text{Cpk} = \frac{\text{Min}(\text{USL} - \text{Average}, \text{Average} - \text{LSL})}{3\sigma}, \qquad (5.13)$$

where USL and LSL denote the upper and lower specification limits, respectively. The limits for BDC, BH, and SS of this process are chosen to be those in Table 5.2, where the USL of SS is selected high enough not to influence the Cpk value. The experimental results are given in Fig. 5.11 for the various ultrasound currents, indicating that for the used capillary-horn combination, the process is within its specifi-

Table 5.2. Specification limits of example process.

	LSL	USL
BDC [µm]	31.6	38.4
BH [µm]	3	13
SS [MPa]	81.1	171.9

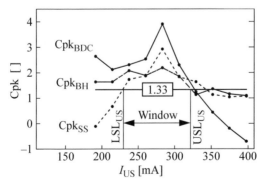

Fig. 5.11. Cpk values versus ultrasound current for 50 µm pitch example process. Averages over 12 chips. One capillary.

cations for ultrasound currents between $LSL_{US} = 229$ mA and $USL_{US} = 323$ mA. The interval centre is $I_C = 276$ mA. As the same types of capillary, wire, and transducer are used for this example process like for the test process, the same ultrasound variations are expected for the example process on several bonders. Other influences of variations such as from the material are not considered. Based on these assumptions, the effects of procedures A-E on the Cpk is evaluated using the standard deviations from Table 5.1, which are shown graphically in Fig. 5.12a.

For each calibration procedure a yield value is evaluated indicating the fraction of capillaries producing bonds within the specifications. The yield value for each calibration procedure is obtained using

$$Y = 100\,\% \cdot \frac{1}{2} \cdot \left[1 + \mathrm{erf}\!\left(\frac{USL - LSL}{2\sqrt{2} \cdot \sigma} \right) \right],$$ (5.14)

where

$$\mathrm{erf}(x) \equiv \frac{2}{\sqrt{\pi}} \cdot \int_0^x e^{-t \cdot t}\,dt$$ (5.15)

and the standard deviations σ from the target values are obtained by multiplying the values given in column 6 of Table 5.1 with the factor $I_C / \overline{I^T}$.

The experimental values of Y are given in column 7 of Table 5.1 and plotted in Fig. 5.12b. The fraction of the capillaries performing within the specifications is lowest for uncalibrated horns (procedure A) and highest when using a calibration after each capillary change (procedure E).

In summary, this investigation uses microsensor measurements, a test process, and an example process to evaluate the potentials of various methods of controlling

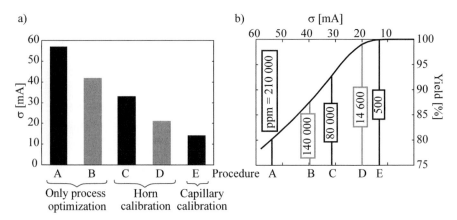

Fig. 5.12. Standard deviations (**a**) and simulated yield (**b**), both for various procedures.

or calibrating the ultrasonic parameter. The yield shows a large sensitivity to the various procedures. The most effective procedure calibrates the ultrasonic tangential force after each capillary change requiring microsensor technology.

The main two obstacles preventing a fast introduction of such a method in production are (a) the possibility that fine pitch capillaries with a delicate bottle neck tip suffer damage from a calibration involving impacts without wire, and (b) the competition with conventional process testing on each bonder, which can be performed even after each capillary change. Such process testing, however, requires material that will be used solely for the ultrasound parameter adjustment bonds, reducing yield and throughput.

Providing an elaborate calibration method for the customer is second to delivering bonders with high uniformity and thus reduced need for calibration. To this end, standard deviations like those in Table 5.1 are continuously sought to be kept ever smaller by improved mechanical and electronic design of the bonders. To accelerate the development of more uniform modules, the following method has been seeing extensive use at ESEC.

Monitoring Bonder Uniformity. The method presented here uses an actual bonded ball to characterize the ultrasound and bond force sensitivity of a bonder, and could also be used for bonder calibration. The parameter profiles are modified by adding calibration segments after the actual ball bond process. Each of these segments has different settings for bond force and ultrasound, as shown in Fig. 5.13. The measured profiles of the transducer current and bond force are shown in Fig. 5.14. A real ball bond is finished after the second impact segment (IS2), and the microsensor force signals (Figs. 5.15a and (b)) are used thereafter for the calibration. The plateau values from the microsensor signals are determined, shown as circles in Figs. 5.15a and (b). These values are plotted versus the nominal parameter values, as shown in Figs. 5.16a and (b). Here, an ESEC specific technical unit,

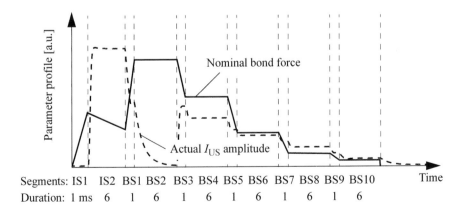

Fig. 5.13. Nominal parameter profiles during ball bond process.

US %, is used as ultrasound parameter, which is proportional to the transducer current. Using real bonds compared to the measurements without wire is an advantage because a larger ultrasound range can be characterized without the need of high normal forces to avoid interfacial friction.

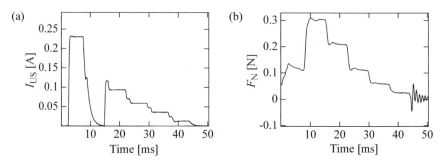

Fig. 5.14. Ultrasound transducer current amplitude (**a**) and bond force (**b**) measured by proximity sensor (horn) during ball bond process.

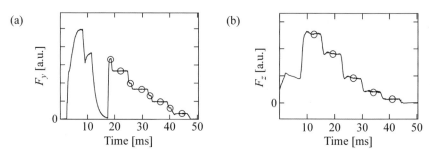

Fig. 5.15. Microsensor force signals during ball bond process. Ultrasound tangential force amplitude (**a**) and bond force (**b**).

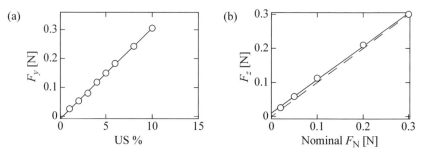

Fig. 5.16. Microsensor forces versus nominal parameters, ultrasound (**a**) and bond force (**b**).

The actual bond force and ultrasonic tangential force delivered to the chip is expected to depend linearly on the bond force and ultrasound parameters, respectively. Therefore, the axis intercept and the slope of the fitted lines obtained from several bonders are sufficient to assess their uniformity. As an example, a number of wire bonders of the same type were characterized in the described way and the resulting calibration curves are shown in Figs. 5.17a and (b). The variations of the curves and thereby the expected process variation from bonder to bonder can be quantified from these data and compared to those of alternative module designs. Obviously, the larger the curves' variation, the smaller the common process window shared by all wire bonders. Based on such results, the best module design with respect to uniformity is determined faster than when using conventional process testing.

5.1.2   Bonding Speed Characterization

Low Frequency Vibrations. In addition to the increased accuracy requirements, the whole bonding process needs to be accelerated to lower the bond production cost, keeping wire bonding economical compared to other bonding technologies. The more units per hour a bonder can put through, the less bonders need to be purchased by the producer.

Speeding-up the process is mainly achieved by stronger motors allowing for faster movement during the wire cycle of a bond. The faster movements also cause more abrupt velocity changes, stronger inertial forces on the bond head, and parasitic vibrations. Even if the bond head is designed to hold the tool and wire in a very stable position relative to the substrate during bonding, a certain amount of such vibrations cannot be avoided at high bonding speeds.

For the process, it is required that the forces produced by parasitic vibrations between tool and substrate are considerably smaller than the ultrasonic and bond forces needed for a successful bond. Excessive vibrations cause excessive deforma-

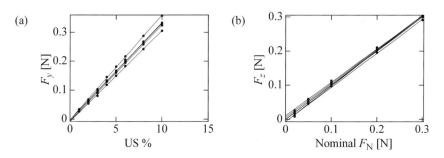

Fig. 5.17. Calibration curves for six wire bonders of the same type, ultrasound (**a**) and bond force (**b**).

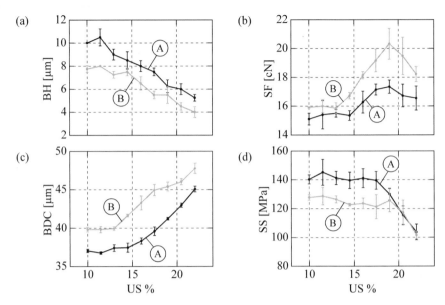

Fig. 5.18. Process results ball height (**a**), shear force (**b**), diameter (**c**), and shear strength (**d**) of reference (A) and test process (B), the latter with a 10 μm horizontal shift. All values are shown as function of the ultrasound settings.

tion of the wire, as is demonstrated by the following experiment. A normal ball bond operation is interrupted after half of its ultrasound time to insert a lateral forth and back movement of the bond head, i.e., a simulation of one horizontal vibration cycle. Then, the ultrasound period is continued to its end. The amplitude of the lateral movement is 0 μm and 10 μm for experiments (A) and (B), respectively, and its cycle time is 60 ms. The measured process responses ball height, shear force, diameter, and strength are plotted in Figs. 5.18a, (b), (c), and (d), respectively. The sample size of each point is 4. Obviously, the lateral shift in this case causes a large additional deformation, resulting in an about 2 μm smaller ball height and an about 3 μm larger ball diameter, in addition to the normal impact force and ultrasound induced deformations. Compared to the reference, the shear force obtained with the shift experiment is larger, too, not because of a larger bond strength but due to the larger deformation. The shift actually reduces the shear strength by almost 20 MPa, which is undesired.

Vertical vibrations are detrimental to ball bonds, too. Usually there is a defined bond force value to achieve the maximum bond strength while keeping the wire deformation below the specified limit. For lower or higher bond forces, a reduced bond strength is observed, and in the case of higher bond forces, excessive deformation occurs. Vertical vibrations cause the bond force to be either higher or lower

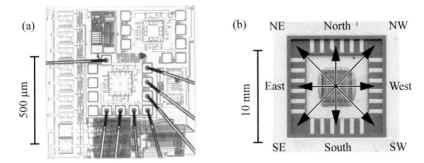

Fig. 5.19. Micrograph of xyz-force sensor with passivated contact zone (**a**), contacted with wire bonds. Test chip in package with test bond directions indicated and named (**b**).

than the optimized value almost all the time during bonding, thus reducing the bond strength and potentially increasing the deformation.

Compared to off-line measurement of process responses, a microsensor method to monitor vibrations is easier to use and results are faster to obtain. For such a method the wire is omitted and the sensor with the passivated contact zone is used. A micrograph of the connected sensor is shown in Fig. 5.19a. Eight test bond directions are used to generate eight different excitations for vibrations. The test bond directions are indicated and named in Fig. 5.19b. The whole test chip setup is shown in Fig. 5.20a. It consists of a custom made heater plate with a PCB attached,

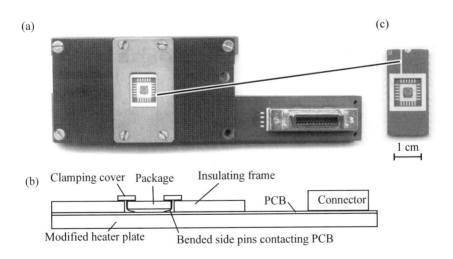

Fig. 5.20. Vibration measurement chip setup. Top view of setup with test chip attached to package (**a**), illustration of cross section of setup (**b**), and top view of ceramics package (**c**).

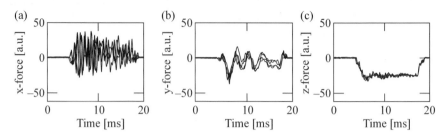

Fig. 5.21. Examples of force vibration signals during test bonds. x- (**a**), y- (**b**), and z-force (**c**). Four signals each graph.

see the illustration in Fig. 5.20b. A ceramic package (CERDIP), as shown in Fig. 5.20c, is used as a chip carrier. The xyz-sensor is contacted with wire bonds to the package leads, and the package pins ensure electrical contact to the PCB.

The test procedure consists of a normal bond procedure including bond head movements but with the ultrasound set to zero in order to allow the undisturbed monitoring of the bond head vibrations. The vibrations are caused by the dynamic movement of the bond head, particularly by the jerk which is generated when the movement reaches the search height where it switches from a high to a low speed. Examples of measured vibration x-, y-, and z-force signals are shown in Figs. 5.21a, (b), and (c), respectively, as obtained with a standard 50 μm fine pitch capillary. To quantify the vibrations, one value is derived from each signal using the following procedure. After applying a 200 Hz high-pass filter to the original signal shown in Fig. 5.22a, an average amplitude is determined from a window section, as illustrated in Fig. 5.22b. This average is calculated by taking the absolute values, then averaging, and finally normalizing the result by multiplying it with 1.571. This

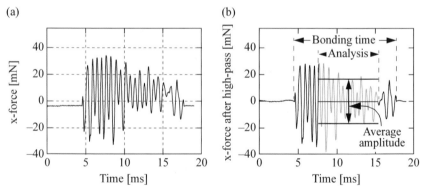

Fig. 5.22. Low frequency vibrations on x-force signal, 10 kHz low-pass (original signal) (**a**) and 0.2 kHz high-pass filter (**b**).

value is chosen in a way that the described averaging procedure would yield 1 for a sine function.

Variation of Search Time. Having demonstrated the disturbing effect of vibrations on the process and how to monitor vibrations with the xyz-force sensor, it remains to demonstrate how the vibrations depend on the bonding speed. A possibility to do this is to vary the search height prior to ball bonding. After the free-air ball formation by EFO, the bond head moves the capillary with maximum speed to the predefined search height above the next ball bond position, where the speed is abruptly changed to the predefined search speed, which is slow enough to allow for reliable touchdown detection. Search height and speed together define the search time. During the search period, the vibrations induced at the search height are settling down to a lower level.

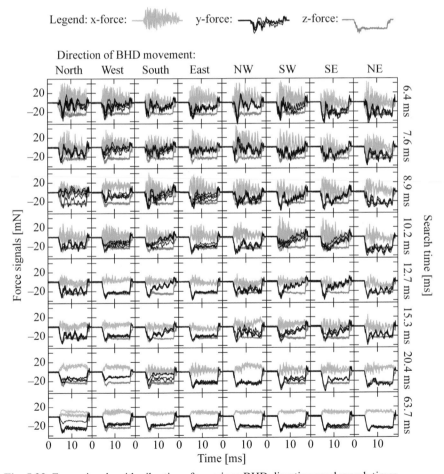

Fig. 5.23. Force signals with vibrations for various BHD directions and search times.

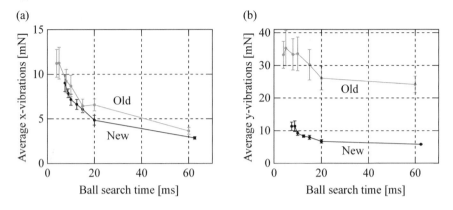

Fig. 5.24. Average amplitude of vibration forces in x- (**a**) and y-direction (**b**), old and new bonder generation as a function of the search time.

For the following example, two wire bonder generations are used. The vibration monitoring method allows to compare the process capabilities of the two bonders. The signals of the first bonder resulting from tests with various search times and the various movement directions are shown in Fig. 5.23. The evaluated average x- and y-force vibration amplitudes as functions of the search time are plotted in Figs. 5.24a and (b), respectively. In this example, the new bonder generation shows an improved vibration behavior and consistently, the process results were found improved, too.

5.1.3   Long-Term Stability

The long term stability of a bonder can be impaired by two effects. One is single dropouts of a parameter during any bond, resulting in a too weak bond and its failure during burn-in testing or later operation. The second is the drift of a parameter during production, causing a drift of the process responses out of the specified limits.

Screening Parameters for Dropouts. During the development of the new bonder generation, the process module stability is being assessed with the monitoring system usually used with the microsensors. However, the microsensors are not needed in this case, as only bonder signals are recorded.

As described in Sect. 3.2, the monitoring system sometimes misses measurements (missed wires) due to the real time limitations of the used PC operating system. The missed wires observed during a test run without bonding real wires are visualized as black spots in Fig. 5.25. About 70 missed wires in a total of 23 595 wires are found, corresponding to 0.3 %. If a parameter dropout exists for such a missed wire, it cannot be detected. Therefore, a 100 % screening is not possible

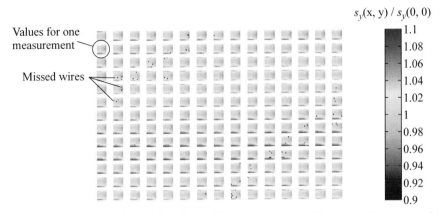

Fig. 5.25. Sensitivity surface of the y-force of 195 measurements. The gray scale corresponds to the normalized signal strength. Black measurements points are missed wires.

without a real time PC operating system. If only 0.3 % of the wires are missed, the probability of detecting parameter dropouts — if they exist — is high.

For the example screening application, a standard bonding process was used. The material was standard QFP 208 leadframes with standard wire bonding test chips. A total of 10 400 wires were bonded on 50 chips and 10 leadframes while the bonder diagnostic signals for ultrasound and bond force were recorded by the monitoring system during the wedge bond.

Figure 5.26 reports the average ultrasound current amplitude of each wire. To compensate for the varying angle between ultrasound oscillation and wire direction, the applied ultrasound strength is intentionally modified as function of wire direction. While the bond force was reliable over the monitored period, two ultrasound dropouts were detected and could not be assigned to an obvious reason. The ultra-

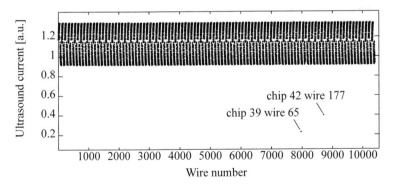

Fig. 5.26. Two wedges bonded with low ultrasound current were detected in more than 10 000 bonded wires.

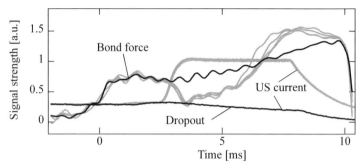

Fig. 5.27. Wedge bond force and ultrasound current signal (black) of a wire with a detected ultrasound dropout. Neighboring wires (gray) show the standard force signal due to the ultrasound enhanced deformation process.

sound currents of the two dropouts (chip 39 / wire 65 and chip 42 / wire 177) are significantly lower than the average. The full signal of the first dropout is shown in Fig. 5.27. This event has impeded a sufficient bond formation and caused a subsequent tail lift-off and therefore a bonding stop. The wire was not deformed enough as is apparent in the corresponding bond force signal. The absence of the usual predetermined breaking point resulted in a misplaced breaking point during tail formation. The second ultrasound dropout was less pronounced than the first and no bonding stoppage was observed. Based on the chip and wire number, the wedge bond was identified after the long term test. Optical inspection revealed an insufficiently deformed wedge that probably would have failed during a reliability test. Together with other considerations, these dropouts found by the screening experiment led to the replacement of the used ultrasound module type by a newly developed type.

Monitoring Parameter Drifts. Drifts of process responses are observed during production and are encountered by, e.g., rigorous 6 σ strategies assuring the desired quality. The drifts can occur due to either material variations or bonder parameter drifts or both. The xyz-force sensor technology can be used to selectively monitor the drifts of the ultrasound and bond force parameters. The setup on the modified heater plate (Fig. 5.20) and the array measurement method (Sect. 4.2.3) are used for this application.

To avoid a too large loss of data, the saddle point evaluation routine was adapted to yield reliable results also when some points are missing. The y- and z-force values were determined about ten times per hour. The hourly averages over a period of about four days are shown for the y- and z-force in Figs. 5.28a and (b), respectively. The drift range of the ultrasound force is about 0.5 % and acceptably low for even the most advanced processes, showing a stable ultrasound module over the four days period. The bond force drift range, however, was about 30 %, a value too high for even standard processes. This result triggered an investigation and corrective

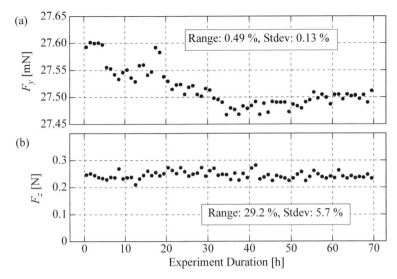

Fig. 5.28. Long term monitoring of **(a)** ultrasonic tangential force and **(b)** bond force, both measured with a microsensor. Each value is an average over one hour. The measurements were performed on a prototype wire bonder.

measures for the bond force module at a very early stage of the new bonder development.

5.2 Ball Bond Process Knowledge

Prerequisite for any bond process investigation is the availability of an inspection method that is highly sensitive to changes in the bond growth process but also provides a high repeatability under constant bonding conditions. Figure 5.29 displays an example of measurements performed with identical bonding parameters demonstrating the excellent repeatability of the measurement system based on the xyz-force sensors. The amplitude and phase of the fundamental and the higher harmonics of the ultrasound forces are related to the wave form shape and are thus changing during the bonding time. Due to the symmetry of the ball bond only odd harmonics are measured. The third harmonic is generally very pronounced. Based on the background knowledge of the wave form and the sequence of the bonding process, the fundamental and the third harmonic amplitudes are good indicators of the duration and strength of the different physical processes taking place during bonding.

The bond process of an Au-Al contact system can be subdivided into phases that are named with the predominant effects *initial sticking* (phase 1), *friction at contact zone* (phase 2), *bond formation* (phase 3), and *friction between ball and capillary* (phase 4). The different effects are often superposed at the transition zones and with

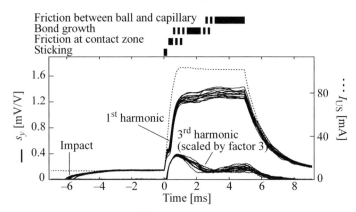

Fig. 5.29. Fundamental and third harmonic of the y-force signal of 14 wires bonded with identical bonding parameters ($F_1 = 180$ mN, $F_N = 100$ mN, $I_{US} = 102$ mA). The chip temperature was 163 ℃. The resulting ball height and diameter are 9.7 ± 1.4 μm and 43.1 ± 7.0 μm, respectively. The shear force is 173 ± 10 mN. The different bonding phases are indicated with the bars on top of the plot. The ultrasound current, a machine parameter, is displayed as a dashed line for reference.

wire deformation effects. Detailed descriptions of the phases 1 to 3 are found in [7, 8, 9].

5.2.1   Friction at Contact Zone

Friction between the Au-ball and the Al-substrate is essential for effective bond growth, and can be identified from the wave form of microsensor measurements [9]. The y-force wave form in Fig. 5.30 is the result of such a friction process. At stage (A), the pad below the contact gets displaced by the motion of the capillary. As long as the force needed for displacing the pad contact zone is below the friction force $F_\mu = \mu F_N$, the capillary and pad are sticking together. The tangential force grows due to further displacement and reaches the limit at stage (B), where friction starts. The capillary and ball are further displaced whereas the pad surface remains at its position. As soon as the capillary reaches its turning point (C), capillary/ball and pad are moving synchronously backward. At the zero crossing of the microsensor signal, the pad has reached its initial position, whereas capillary and ball are still displaced (a result of the preceding relative movement). The same process will take place for the capillary movement in the opposite direction. This stick-slip motion results in a relative movement between gold ball and pad. Thus, the position of the capillary tip relative to the chip coordinate system is described by two vertically shifted sine curves. Calculations of the friction energy are reported in [5, 10].

Some recorded friction wave forms show an asymmetry between the positive and negative half wave. This asymmetry can be explained by an additional normal force oscillation at ultrasound frequency. Figure 5.30 shows the measured z-force oscilla-

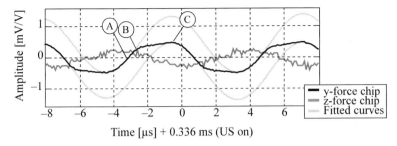

Time [μs] + 0.336 ms (US on)

Fig. 5.30. Measured wave forms of the y- and z-force signal during friction at ball-pad contact. For explanation of the markers (A) to (C) refer to text. The lightly grey sine functions are fitted to the y-force signal during the sticking phases. A harmonic analysis of this measured y-force will primarily yield odd harmonics (see Fig. 5.29).

tions with a phase difference to the y-force, which is typical for the used transducer system. A mechanical cross-talk between y-force and z-force could be excluded by experiments on passivated sensors if the ball bonds are sufficiently centered.

During bond growth, the maximal tangential force increases due to the improved adhesion between gold ball and pad.

5.2.2   Friction Between Ball and Capillary

Although bonding takes place at the contact zone, the maximum tangential force that can be applied to the contact zone is limited. Fig. 5.31 shows the measured tangential ultrasound force saturation of a 60 μm pad pitch process. The impact force and bond force parameters are 180 mN and 100 mN, respectively. In combination with a free transducer amplitude of 249 nm (I_{US} = 102 mA), these parameters yield balls within the process specifications. In the following, the origin of the tangential force saturation is identified by interpretation of the wave form shape.

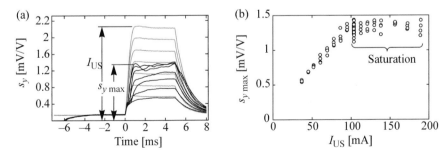

Fig. 5.31. Saturation of s_y for increasing ultrasound amplitude (**a**). The ultrasound current is scaled to match with s_y before US on. Measured maximal y-force sensor signal $s_{y\,max}$ as a function of the ultrasound current (**b**). (F_N = 100 mN, F_I = 180 mN).

The friction process (friction at contact zone) during the first phase of bonding results in a peak in the 3rd harmonic of the y-force signal. The bond growth between pad and ball results in a decrease in the 3rd harmonic and an increase in the first harmonic as shown in Fig. 5.32a. If no bond growth is possible (e.g. passivated pad), the 3rd and 1st harmonic remain at a constant (force dependent) level. If only this friction process took place, the third harmonic would be zero at the end of the bond growth. But the third harmonic does only vanish for bond parameters that inhibit friction during the whole ultrasound bonding process. This is e.g. the case for too low ultrasound amplitudes. For normal bond parameters, the third harmonic approaches a constant value for high bond force and low ultrasound values or grows once again for high ultrasound amplitudes as shown in Fig. 5.32b. A close look at the wave form shows that the process which causes this increase behaves differently than the friction process at the contact zone where friction between gold ball and pad takes place at maximum tangential force values. Figure 5.33 shows a typical wave form during the second increase of the third harmonic. The y-force wave form shows kinks (A) just after the zero crossing of the y-force (see Fig. 5.33). The distance from zero is bond force dependent which is pointing towards friction processes. The strength of this process depends highly on the applied bond force and ultrasound amplitude. Low bond force in conjunction with high ultrasound amplitude provoke strong kinks. For these process parameters, vertical ball deformation is absent. Furthermore, the wave form of the z-force oscillation changes significantly.

All these characteristic features in the force signals can be explained by friction processes between the capillary and the ball. During the 2nd phase of the bond process, the friction at the contact zone is limiting the maximum force acting on the gold ball. As friction at the contact zone is inhibited due to the bond growth, strong tangential forces are applied on the ball. For the used process and bonding tool, ultrasound amplitudes above about 244 nm (100 mA) would exceed the maximum attainable tangential force (fully bonded wire of a 60 μm pad pitch process) if no other mechanism lowering the force were present. The tangential force is high

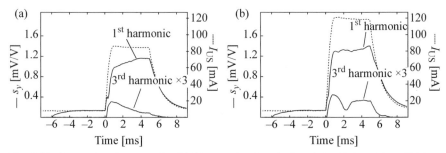

Fig. 5.32. Comparison between the third harmonic of a wire bonded with low (**a**) and high (**b**) ultrasound current. The amplitude of the third harmonic is scaled by a factor of 3. (F_N = 100 mN, F_I = 180 mN).

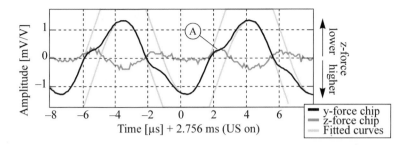

Fig. 5.33. Microsensor signal wave forms during friction between capillary and ball. A positive z-force sensor voltage corresponds to a decrease in the force.

enough to initiate plastic deformation of the gold ball. For a large bond force and moderate ultrasound amplitude, the ball is severely deformed in vertical direction. If the ultrasound amplitude is high, a large tangential force is acting on the neck of the ball. A deformation of the neck results in a loosened grip between capillary and ball and relative motion between ball and capillary occurs. The kink in the y-force signal is thus due to the change of the ball side contacting the capillary.

The z-force signal mainly consists of a first harmonic during phases 1 to 3. As soon as the kink in the y-force signal appears, the wave form of the z-force signal changes fundamentally. As the kinks in the y-force become visible, z-force depressions appear synchronously in time with the y-force kinks (see Fig. 5.33). This results in a strong second harmonic. As the capillary-horn system behaves inert to position changes at high frequencies, a sudden z-position change of the capillary tip will be mainly seen in a change of the bond force. The capillary-horn system can not follow fast enough due to its mass. By considering this behavior the z-force wave form can be explained as a change in z-position due to the change of the ball side contacting the capillary. The z-force wave form is thus a combination of the ultrasound caused z-force oscillation at ultrasound frequency and the z-force depression peaks.

To get a better understanding of this process, a mechanical ANSYS® FE model is employed. Figure 5.34a shows the mesh used for the simulation. As relative motion between the rigid capillary tip and the ball is allowed and the gold is treated as a plastic material, only one symmetry plane exists. As the simulation targets the understanding of the observed reaction forces and displacements, it is based on a predeformed ball geometry. Therefore, the resulting stress field distribution in the ball is only an approximative solution as strains resulting from impact deformation are neglected. The gold ball model is composed of a mapped mesh of cubic SOLID95 elements. The capillary tip is shaped with rigid mesh elements attached to a pivot node. The contact between capillary and ball is formed with CONTACT174 and TARGET170 elements.

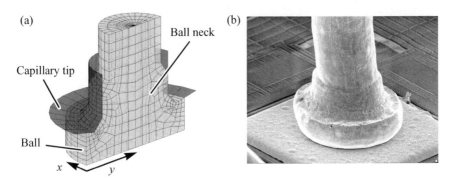

Fig. 5.34. Mesh used for simulation (**a**) and bonded ball on a Au-plated pad (**b**). The mesh of the capillary tip surface is rigid. Relative movement between capillary tip and ball is enabled by contact elements.

The contact zone displacement at the pad is set to zero (no pad-ball friction). The time evolution of the simulation is controlled by applying a z-force and a displacement in y-direction on the pivot node of the rigid capillary. The displacement in x-direction and the rotation of the rigid capillary are set to zero. The profile of the applied bond force and displacement is shown with dashed lines in Fig. 5.35. The force value is the force that acts on the full capillary geometry.

As extended yield material parameters for the gold wire were unavailable, a simple bilinear kinematic hardening model is applied for the simulation [11]. The stress-strain relation of this simple plasticity model is shown in Fig. 5.36. A yield stress of 140 MPa is chosen, the slope in the plastic region of the stress-strain rela-

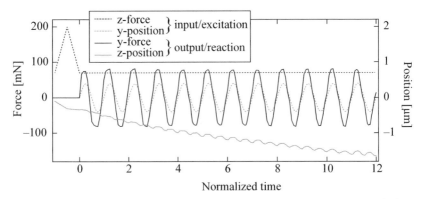

Fig. 5.35. Simulated time evolution of the forces and displacements under an applied normal force (z-force) and oscillating y-movement (y-position) of the capillary. For further detail about the model parameters refer to the text. Dashed and solid lines are excitation and reaction signals, respectively.

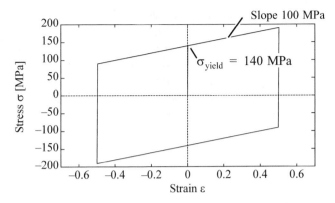

Fig. 5.36. Bilinear kinematic hardening model used for the gold.

tion is 100 MPa. A more accurate parameter model would request the simulation of the entire deformation process during the impact to account for the strain hardening as a result of the strong deformations.

Figure 5.35 shows the reaction forces and displacements of a simulation of 12 periods of the ultrasound oscillation. The simulation of the plasticity and the contact results in time consuming load step calculations combined with iterations as a result of contact penetration. The z-position in Fig. 5.35 indicates that the deformation of the ball saturates after more than 12 oscillations. A close look at the y-force and the z-position (see Fig. 5.37) shows similar behavior to the wave form observed in mea- surements. As the simulation is not yet stationary, deformation processes at the maximum displacement amplitude (A) are still observable. The kink (B) is the result of the relative motion between capillary and ball and is similar to the mea- surement in Fig. 5.33. Figure 5.38 shows the simulated gap between the capillary and the ball at maximum displacement.

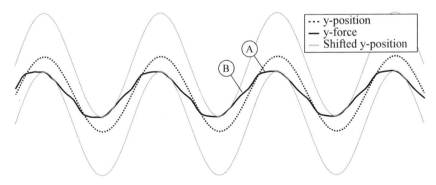

Fig. 5.37. Detail of the numerical excitation (dashed line) and reaction forces (black line).

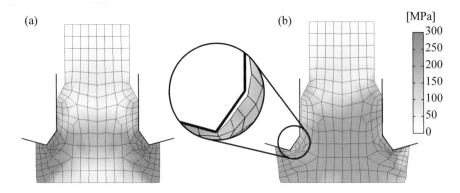

Fig. 5.38. Von Mises stress due to an applied bond force $F_N = 200$ mN (**a**) and for a combination of normal force $F_N = 90$ mN and tangential displacement amplitude $D_y = 400$ nm of the capillary tip (**b**). The detail shows the gap between gold ball and the capillary for maximal capillary displacement.

5.2.3   Deformation

The bond process window is not only defined by sufficient contact adhesion but also by a controlled deformation of the ball. This is especially important for small pad pitches as overdeformed balls can result in a short-circuit of neighboring pads and diminish the loop height consistency. Ball deformation is a result of the impact force and/or the combination of an applied normal force and an ultrasound tangential force during the ultrasound bonding time. To distinguish the two deformation phases, the latter will be referred to as *ultrasound enhanced deformation*. Due to the reduced symmetry of the ultrasound enhanced ball deformation and its high sensitivity to process variations, this deformation is often avoided by choosing a high impact force. Thus, a simple criterion for the onset of deformation under ultrasound stress would be beneficial for process optimization. A suitable criterion will be derived in the following section.

The onset of plastic deformation of a material is described by a yield criterion [12]

$$f(\sigma) > \sigma_{\text{yield}},\tag{5.16}$$

whereas the functional $f(\sigma)$ is a function of the deviatoric stresses $s_{ij} = \sigma_{ij} - \delta_{ij}\sigma$. It is invariant under coordinate transformation. The Von Mises stress

$$\sigma_{\text{VonMises}} = \sqrt{\frac{1}{2}[(\sigma_1 - \sigma_2)^2 + (\sigma_1 - \sigma_3)^2 + (\sigma_2 - \sigma_3)^2]}\tag{5.17}$$

is based on the second principal invariant of the deviatoric stress. The stress components σ_i ($i = \{1,2,3\}$) are the principal stresses. The yield stress σ_{yield} is defined as the point of 0.2 % deviation from the linear stress strain relation.

The complex geometry of the deformed ball results in a non-uniform stress field distribution in the ball. Therefore, a rigorous description needs FEM simulations for calculation of the principal stress values. Such FEM based deformation models generally give qualitative data but it is difficult to extract significant process relevant data as the whole bond parameter space has to be considered [13, 14]. A rigorous plane strain solution with inclusion of thermal and strain hardening effects is found in [15].

The combination of the applied normal force and the shearing traction results in dominating stress fields σ_{zz} and σ_{yz}.

$$\sigma \approx \begin{bmatrix} 0 & 0 & 0 \\ 0 & 0 & \sigma_{yz} \\ 0 & \sigma_{zy} & \sigma_{zz} \end{bmatrix} \; ; \; \sigma_{yz} = \sigma_{zy} \, . \tag{5.18}$$

Substituting the eigenvalues of the stress matrix (5.18) into equation (5.17) yields the Von Mises stress

$$\sigma_{VonMises} = \sqrt{\sigma_{zz}^2 + 3\sigma_{yz}^2} \, . \tag{5.19}$$

The stress fields are related to the applied force by integration over the cross sectional area C of the ball.

$$F_j = \int_C \sigma_{jz} df \, . \tag{5.20}$$

The assumption of a homogeneous stress field distribution yields the stress fields

$$\sigma_{jz} \approx \frac{F_j}{A} \tag{5.21}$$

for a known force and cross sectional area A. The stress field σ_{zz} is generally underestimated by equation (5.21) as the stress field is concentrated at the ring shaped contact between capillary tip and ball.

The yield stress limit in equation (5.16) is defined by the point in the stress-strain relation that results in a 0.2 % strain offset from the ideal elastic stress-strain function. This convention is used to get a reliable yield criterion even though the elastic limit will be reached at a lower stress value. Measurements of the stress-strain relation of gold wires are found in [16] for different temperatures. For large strain values the functional dependence may be approximated by a new linear relationship with a flat slope. A schematic stress-strain relation of Au is shown in Fig. 5.39. Dur-

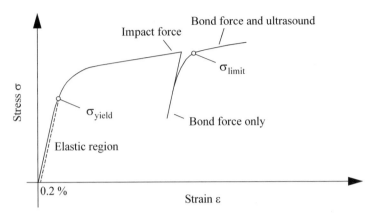

Fig. 5.39. Schematic drawing of the stress-strain relation during the bond formation for Au. The critical stress value σ_{limit} for gross plastic deformation after the impact is higher as the yield stress σ_{yield} defined by 0.2 % plastic strain offset from the ideal elastic relation.

ing the ball bond process the impact deformation will result in large plastic strain values. As the impact deformation is not studied in this work, it is included as a strain-hardening effect.

After the impact, the normal force is reduced to the bond force value. This results in a stress value that is considerably lower than the yield criterion. If ultrasound is applied, the combination of the tangential and normal force may again cross the yield criterion causing ultrasound enhanced deformation. This reloading will result in a higher plastic limit, also known as *strain aging* [12, 17]. The ultrasound enhanced deformation process will only be of significance if a large plastic strain is observable. Under the condition of absent ultrasound softening, this limit will approximately correspond to the impact stress value. To take account of this gross deformation effects, a modified yield criterion σ_{limit} is used for subsequent considerations. This value will vary for different impact forces and ball volumes as different levels of strain-hardening are achieved. The plastic properties of the Au after ball formation are difficult to obtain.

Table 5.3 lists the properties of the used gold wire AW-14. The plastic properties of AW-14 wires are not available to the public. Table 5.4 lists the plastic properties of high purity gold wires and melted free air balls (FAB). Large variations are observable as a result of different impurity concentrations, preparation methods and measurement methods used for extraction of the yield strength. The wire drawing process results in strain hardening that causes a high yield strength. Subsequent annealing processes (e.g. storage at elevated temperature) significantly reduce the yield strength. The melting and recrystallization process during the free air ball formation changes the structural property of the gold ball as well. Based on Vickers indentation tests, a yield strength σ_{yield} = 132 MPa is reported in [18] for free air

Table 5.3. Properties of the gold wire AW-14.

Property of gold wire AW-14	Value	Ref.
Purity [%]	> 99.99	[20]
Young's modulus [GPa]	79	[20]
Poison ratio [ ]	0.42	[21]
Tensile strength [MPa]	180 - 260 225 ± 15	specification measured

balls, whereas the yield strength of the original wire is 183 MPa. Caution is needed when comparing values extracted from indentation tests with values extracted from tensile tests. The stress-strain curve of fully annealed high purity gold is found in [19].

Table 5.4. Plastic properties of high purity gold.

Plastic properties of gold with 99.99 % purity		Temp. [°C]	Yield strength [MPa]	Tensile strength [MPa]	Ref.
Wire	drawn, not annealed, 25 μm diameter, 1 ppm Be	R.T.	235 [a]	245	[22]
Wire	drawn, annealed, 25 μm diameter	R.T.	181 [a]	203	[22]
Wire	6 ppm Ca, ~1 ppm Ag Cu Fe Mg, 25 μm diameter	200 160	152 [a] 180 [a]	214 220	[16]
Wire	drawn, annealed 300 °C 0.6 s, 3 ppm Be, 25 μm diameter	R.T.	180.6 [a]	217.9	[23]
Wire	drawn, annealed 300 °C 10 min, 3 ppm Be, 25 μm diameter	R.T.	45.9 [b]	98.6	[23]
FAB	75 μm diameter	R.T.	132 [b]	—	[18]
Wire	annealed, 500 °C, 1 mm diameter	R.T.	86.48 [a]	—	[19]

(a) *Based on tensile tests.*
(b) *Based on Vickers indentation tests.*

The plastic deformation during the impact phase causes strain hardening. The amount of strain hardening depends on the impact condition and the degree of ball deformation. It follows from measurements that the maximum attainable shear strength (at 25 °C) of bonded balls is about 130 ± 10 MPa (see Fig. 5.46). Changes of the impact conditions can affect this value.

The shear σ_{yz} strength can be used to estimate the yield criterion of a deformed ball

$$\sigma_{limit} = \sigma_{yz}\sqrt{3} \approx 225\,\text{MPa} \pm 17\,\text{MPa}. \tag{5.22}$$

The estimation of the yield criterion based on shear test measurements is approximative as the shear test itself causes strain hardening. During the impact phase the tangential force is small in comparison to the normal force and equation (5.19) can be simplified to $\sigma_{zz} = \sigma_{limit}$. Thus, a calculation of the impact force under the assumption of a homogeneous stress distribution σ_{zz} and neglecting of other stress components yields an impact force $F_N = 372$ mN for balls with a contact zone diameter of 45.9 µm. The cross at 363 ± 13 mN in Fig. 5.40 marks the measured impact for balls with this contact zone diameter. Each circle in Fig. 5.40 stands for a bonded ball. The bond force and the tangential force are measured at the start of the ultrasound enhanced deformation. The time point of the deformation start is extracted from the bond force and the z-position signal of the bond head. The shaded areas in the parameter space are not accessible for measurement (see Sect. 5.2.5). Based on equation (5.21) the limit for deformation can be calculated as function of the applied normal force and tangential force. The elliptical contour lines in Fig. 5.40 mark force combinations that cause equal Von Mises stress. The measure-

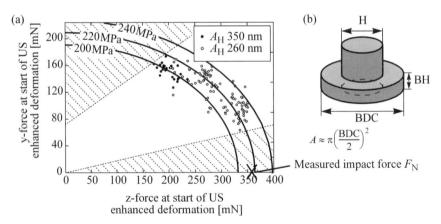

Fig. 5.40. Measured combinations of z-forces and y-forces at the start of the ultrasound enhanced deformation (**a**) for varied ultrasound amplitude and bond force, adopted from [27]. The dashed areas mark regions that are not measurable. Ball geometry (**b**) used for the simple deformation model. (Sensor sensitivity $g_{yy} = 9.28$ mV/V/N).

ments are performed with a relatively large impact force. The large strain hardening results in a relatively large Von Mises stress that is needed to start ultrasound enhanced deformation. To account for the differences in the ultrasound caused strain hardening, the yield limit is extracted from the measured impact force for subsequent measurements.

The above consideration is based on superposition of the ultrasound stress field with the bond force. The influence of ultrasound vibrations on the plastic deformation of material is disputed since Langenecker's work [24] in 1955. An extensive discussion of such effects is found in [25]. Heating effects due to ultrasound friction and dissipation processes are neglected in this model. Measurements of the temperature increase due to the ultrasound are found in [7].

5.2.4   Comparison with Other Measurement Methods

The microsensor measures the force that acts on the pad of the die. The friction processes that occur during bonding also have an influence on the capillary vibration behavior, as the boundary condition at the capillary tip changes during the bond process. Laser vibrometers have been extensively used for wedge bond process investigations [26], where relative motion between wire and pad was detected. As the adjustment of the laser spot is difficult, only single measurements can be performed. Figure 5.41 shows a measurement with simultaneously recorded microsensor signals and laser vibrometer measurement (LIM). A small ultrasound current amplitude is present before the ultrasound bonding start (US on) at $t = 0$. This minimal ultrasound current amplitude is needed to preserve the phase locked state of the used ultrasound system. It causes a vibration of the capillary as observed in the first harmonic of the laser vibrometer signal.

The laser spot is focused on the bottle-neck of the capillary just above the tip. The size of the laser spot on the capillary and mechanical relative movements between bond head and laser unit impede an exact localization of the laser vibrometer measurement position. Therefore, the measured displacement amplitude value can not be related to bonding parameters or sensor signals. However, the time evolution of the LIM signal includes information about the strength of the different physical processes.

As the ultrasound is turned on at time zero, the laser vibrometer amplitude also rises. However, due to various physical processes the 1st harmonic of the laser vibrometer amplitude profile diverges from the ultrasound current profile. The reason for the different behavior is the different distance between measurement position and the bond contact. On the ultrasound current, the influence of the feedback from the boundary condition at the capillary tip is below the measurement limit. The maximum LIM displacement amplitude of the 1st harmonic is 43 nm. A maximal value of 8 nm was found for the 3rd harmonic. A correlation between the third harmonic of the microsensor y-force signal and the 3rd harmonic of the vibrometer is clearly visible in Fig. 5.41. As the capillary tip becomes clamped after impact (A), the laser vibrometer amplitude decreases. If the tangential force (low-pass y-force signal) is high enough (B), sliding between ball and pad results in an almost freely vibrating capillary (C). This is also the case during the ultrasound induced friction period (D). The microsensor measures the force acting on the pad. Thus, the bond growth initiated by the friction process between ball and pad causes an increase of the force that can be applied to the pad as observed in the 1st harmonic of the microsensor y-force signal. This is opposite to the behavior seen in the 1st harmonic of the LIM signal. Bond growth at the contact zone results in a decrease of the capillary tip vibration amplitude.

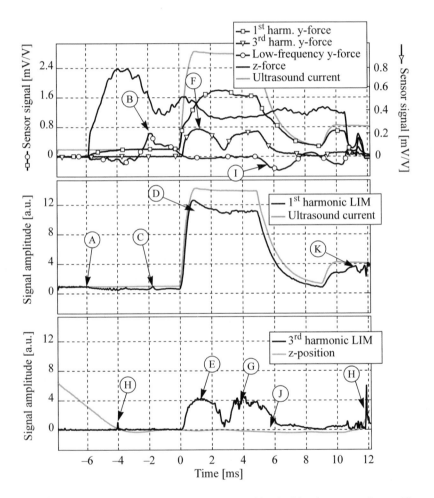

Fig. 5.41. Laser vibrometer measurement (LIM) combined with microsensor forces. The sensor has a pad opening of 65 µm (small Al sensor). The machine feature 'bond step wise' was used to perform the measurement. The laser spot covered the bottleneck of the capillary. Measurements at the tip were not possible because of a too large laser spot diameter. Details A to K are explained in the text. Measured on a WB 3088iP.

A strong third harmonic is observable in the laser vibrometer (E) and the microsensor y-force signal (F). Furthermore, the reascend of the third harmonic also appears in the laser vibrometer signal (G). The two peaks (H) are distortions and are not to be interpreted. As soon as the bond adheres to the pad, low-frequency y-forces (I) can be transferred to the pad. The force acting on the gold ball is high enough to initiate ultrasound enhanced deformation (J) which can be seen with decrease of the z-position. During the lift-off phase the capillary is vibrating freely and the LIM signal follows the ultrasound current (K). The recorded measurement was bonded stepwise and is therefore not representative for normally bonded wires.

If the laser spot is directed at higher positions of the capillary, e.g. the main taper angle, the higher harmonics in the laser vibrometer signal are below the measurement limit. This observation is in accordance with simulation results of the displacement profile of the free and clamped capillary as e.g. shown in Fig. 3.13. Above about 800 µm from tip the difference in the displacement profile between a free and clamped capillary is minimal.

The amplitude of the second harmonic is below the measurement limit. The absence of all even harmonics is apparent, as it follows directly from the symmetry properties of the ball bond. As the geometry of wedge bonds break this symmetry even harmonics may be observed during the bonding of wedge bonds. Actually, signatures of second harmonics were measured with PZT stress sensors mounted on the transducer horn of a wedge-wedge bonder and a correlation between bond quality and the finger-print of the measured 2nd harmonic is reported in [28].

5.2.5   Bond Process Parameter Window

In the previous sections, the consequences of the physical processes on the bonding behavior were explored for fixed bond parameters. Relevant for wire bond process optimization is the knowledge of the influence of different machine settings to achieve a robust and reliable bond quality. The following section elaborates on the connection between the physical processes that affect the bond process and the machine settings. The relation is exemplarily explained with ball bond process parameter windows of a 100 µm and a 60 µm pad pitch process.

Figure 5.42 shows bond process parameter windows for a varied bond force and ultrasound oscillation amplitude. The measurement is based on process conditions of a 100 µm pad pitch process. The measured impact force is 425 ± 15 mN. The chip temperature is 163 °C. The used capillary SBNE-38BD-C-1/16-XL-20MTA from SPT Roth Ltd, Lyss, Switzerland, has a hole diameter of 38 ± 0.5 µm, 47.5 ± 0.5 µm chamfer diameter, and a 11° face angle. A gold wire AW-14 with 33 µm diameter is used. Optimal bonding parameters for this process are found around 200 mN bond force and 252 mA ultrasound current amplitude, resulting in 18.9 µm ball height (BH) and 57.1 µm ball diameter at the capillary imprint (BDC). The impact deformation results in 19.9 ± 1.1 µm ball height and 55.2 ± 0.3 µm ball diameter. In total 624 wires were bonded with varying ultrasound current and bond force and the forces were recorded in-situ with the microsensor system. After bonding, the shear force (SF) was measured off-line with a shear tester. The deformation during the ultrasound bonding time (US enhanced deformation) is extracted from the bond head z-axis position measurement system. It is the difference of the position at the start and the end of the ultrasound bonding phase.

The contour plots in Figs. 5.42a and (c) are the bond process parameter windows as a function of the applied bond force and ultrasound current. The ultrasound enhanced deformation and the shear force are parameters used for conventional bond process optimization. The contour plots (b) and (d) show the same data as a function of the physical quantities acting on the contact zone. The dashed area marks zones that are not accessible to the measurement. At low ultrasound current amplitudes ball non-sticks are observed, i.e. the adhesion between ball and pad is too low and the ball will thus not stick to the pad after bonding. This ball non-stick zone is delimited by a sharp line (B) denoted *line of friction*. The slope (ratio of y-force to bond force) is $\mu_0 = 0.38$ for Au-Al contact systems at room temperature [5]. It is found from microsensor measurements that below this line gross sliding between ball and pad is absent, resulting in ball non-sticks for moderate bond temperatures (i.e. not allowing for thermo-compression bonds). For low bond forces, the ball non-stick limit deviates from the line of friction due to friction effects between ball and capillary. The ball gets stuck to the capillary during impact defor-

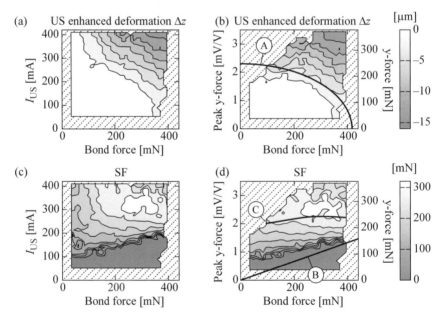

Fig. 5.42. Shear force (SF) and ultrasound enhanced deformation contour plots as function of the bond force and ultrasound current amplitude or measured y-force amplitude. The dashed area marks zones that are not accessible to the measurement for physical reasons or machine limitations. For discussion of the deformation line (A), friction line (B), and gold shear line (C) refer to text. The bond force is measured with the force sensor of the wire bonder machine. (Sensor sensitivity $g_{yy} = 9.28$ mV/V/N).

mation and is torn off the pad during lift-off at the end of the bonding process if the adhesion is small. The region of lowest bond forces below about 40 mN is not accessible with the used machine model. Bond forces that are higher than the impact force are of no interest as they result in excessive deformation during ultrasound bonding. The contour lines of both the shear force and the ultrasound enhanced deformation show a hyperbolic characteristic as function of the applied bond force and ultrasound amplitude with exception of the ball non-stick region. The distance between neighbouring contour lines in Fig. 5.42 is 40 mN and 2 μm for the shear force and ultrasound enhanced deformation, respectively. The contact adhesion of balls above line (C) is large enough to get gold shear (a part of the gold ball remains on the pad after the shear test). Due to friction, the tangential ultrasound force amplitude acting on the ball is not directly related to the ultrasound amplitude. The peak value s_y^{peak} of the ultrasound y-force signal at US off is extracted from the measurement and used as measure for the tangential force acting on the ball. The SF and ultrasound enhanced deformation as function of the applied bond force and the peak y-force is displayed on Figs. 5.42b and (d), respectively. As the values are extracted at the end of the bonding time, friction at the contact zone has ended for most of the used settings. The tangential force saturation as a result of friction between ball and capillary limits the maximal force amplitude for low bond forces. The equipotential lines of the measured SF become in first order independent of the applied bond force (with the ball non-stick region excluded).

The following considerations are only valid for small deformations based on the assumptions of a constant ball geometry. Under the assumptions discussed in Sect. 5.2.3, the deformation line can be drawn in the graph by knowing the impact force. The line of deformation (A) in Fig. 5.42 is deduced from the Von Mises yield criterion defined by equation (5.19). The force that the capillary tip is transferring to the ball under clamped condition is directly proportional to the ultrasound current (horn tip amplitude) and can be calculated with the capillary model of Sect. 3.2. Thus, during the transient when the ultrasound is switched off and the vibration amplitude is decreasing, the ratio between ultrasound current and measured y-force sensor signal should be constant for bonded balls. To test this prediction on the measurement, a scaling value between y-force and ultrasound current (7.46 mV/V/A for the data shown in Fig. 5.43) is extracted from non-stick bonds (no friction and deformation). The ultrasound current value is translated into a theoretical y-force with this factor for each measurement. The force that is derived from the ultrasound current and the measured force should thus be identical for measurements that comply with the prediction. The lines joining both force values in Fig. 5.43 indicate a deviation from this constant scaling value. The higher y-force microsensor signal for heavily deformed wires (A) is explained in Sect. 3.2. The friction between ball and capillary in region (C) results in large deviations at small bond forces and high ultrasound values. In fact, all measurements in the region (C) are relocated to line (E). All measurements consistently show an offset of this demarcating line. Figure 5.44 schematically summarizes the different regions in the ultrasound current / bond force process parameter window.

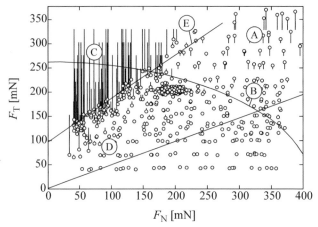

Fig. 5.43. Comparison between forces measured by the microsensor (circles) and forces calculated from the ultrasound current values (upper end of the vertical lines), joined to circles by vertical lines. Values are identical if no line is visible. The scaling factor between microsensor measurement and setting value of the current is 7.46 mV/V/A. Deviations in the region (A) are explained by a force increase at the contact zone due to ball deformation (see Sect. 5.2.3). The Line (B) is the ball deformation line prediction extracted from the impact value 430 mN. Deviations in region (C) are explained by neck deformation that limits the maximal force transfer. If friction at ball-pad contact zone is present deviations in region (D) are observable. (Sensor sensitivity $g_{yy} = 9.28$ mV/V/N).

The influence of the physical processes was verified by recording bond parameter windows for modified process conditions. Figure 5.45 shows the bond process

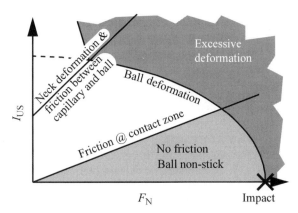

Fig. 5.44. Schematic of the demarcating lines in the process parameter window. The ultrasound force (ordinate) stands for the maximal force that the capillary can exert on the ball under absence of other force limiting effects.

parameter windows based on settings of a 60 μm pad pitch process. The chip temperature is 163 °C. The used capillary SBNE-28ZA-AZM-1/16-XL-50MTA from SPT Roth Ltd, Lyss, Switzerland, has a 28 ± 0.5 μm hole diameter, 35.5 ± 0.5 μm chamfer diameter, and a 11° face angle. A gold wire AW-14 with 22 μm diameter is used. The measured impact force is 204 ± 3 mN. The impact deformation results in 11.9 ± 1.1 μm ball height and 41.1 ± 0.5 μm ball diameter. Optimal bonding parameters for this process are found around 90 mN bond force and 100 mA ultrasound current amplitude, resulting in 9.7 μm ball height (BH) and 43.1 μm ball diameter at the capillary imprint (BDC). The contour plots on the left of Fig. 5.45 are the bond process parameter window as a function of the applied bond force and ultrasound current. The ball non-stick zone is delimited by the *line of friction* (B). The distance between neighboring contour lines in Fig. 5.45 is 20 mN and 1.5 μm for the shear force and ultrasound enhanced deformation, respectively. Similar to the 100 μm pad pitch process, the friction between ball and capillary limits the maximal ultrasound force that can be transferred for low bond forces.

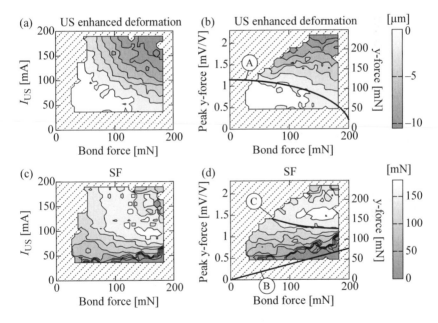

Fig. 5.45. Shear force and ultrasound enhanced deformation contour plots as function of the bond force and ultrasound current amplitude or measured y-force amplitude. The values are measured at the end of the ultrasound time. For discussion of the deformation line (A), friction line (B), and gold shear line (C) refer to text. The dashed area marks zones that are not accessible to the measurement for physical reasons or machine limitations. The bond force is measured with the force sensor of the wire bonder machine. (Sensor sensitivity $g_{yy} = 9.28$ mV/V/N).

To control the quality of a wire bond production line, both the shear force and the ball geometry are routinely measured off-line. Information about the geometry can be extracted on-line from the z-position signal of the bonding tool. However, the ball deformation does not give full information about the strength of the bond. A good correlation between the peak value of the microsensor y-force signal and the conventional shear test quality parameter was found for Al-Au wire bonds, as shown in Fig. 5.46. There is no correlation for balls with the failure mode 'gold shear', as the conventional shear test does no longer characterize the adhesion at the contact zone. Ball non-sticks can be detected in the microsensor force signal, as friction is absent during such bonds. A deviation is observable for bonds with low shear forces. For these bonds friction endures till the end of the ultrasound bonding time. As this tangential friction force is bond force dependent, the microsensor senses a larger force than the off-line measured shear force after bond force removal. This is no more the case if the tangential force is dominated by bonded contact zones that have to be broken. The measurements have been repeated at different substrate temperature, for different capillary types, and for different wire diameters. A linear dependence between shear force and peak y-force comparable to those in Figs. 5.46 and 5.47 was found for all measurements.

Measurements at 35 °C substrate temperature are shown in Fig. 5.47. As the bond formation process is generally slower at low substrate temperatures, balls bonded with identical parameters as in Fig. 5.46 show lower shear force values. Deviations are visible for balls with low shear force that are bonded with high bond force. The scaling factor between peak y-force and shear force is 10.5 mV/V/N. To summarize, the following observations are made:

- The shear force is independent of the applied bond force for sufficiently strong bonded balls without ultrasound enhanced deformation.

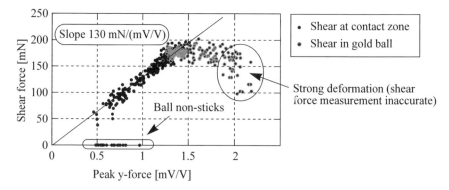

Fig. 5.46. Correlation between standard shear force and ultrasound force measured by the microsensor for varied bond force and ultrasound current amplitude. (F_I = 180 mN, F_N = 50 mN ... 162 mN, I_{US} = 20.4 mA ... 204 mA, substrate temperature T_s = 163 °C, ultrasound bonding time t_{US} = 5 ms).

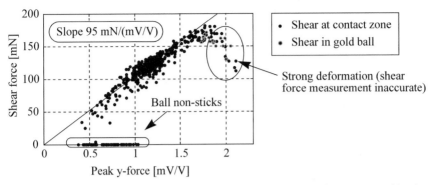

Fig. 5.47. Correlation between standard shear force and ultrasound force measured by the microsensor for varied bond force and ultrasound current amplitude. ($F_I = 180$ mN, $F_N = 50$ mN ... 175 mN, $I_{US} = 20.4$ mA ... 204 mA, substrate temperature $T_s = 35$ °C, ultrasound bonding time $t_{US} = 5$ ms).

- Bonds with low shear force are overestimated by the peak y-force criterion. The deviation from the linear fit is highest for large bond forces.
- Negligible bond growth takes place if friction at the contact zone is absent for bonds of Au-Al contacts.
- The peak y-force measured by the microsensor matches the shear force for contacts bonded at room temperature, i.e. a fit between the peak y-force and the shear force yields a slope of value one.
- The shear force (measured at room temperature) is higher than the peak y-force measured during the bond process if the substrate temperature is above room temperature during bonding.
- Measurements indicate that the correlation holds for balls bonded with varied bonding times. However, for very short bonding times (below e.g. 3 ms) the y-force of the stick-slip transition point has to be determined to achieve a correlation (see Figs 5.48 and 5.49). With progression in the bond growth the difference between the y-force at the stick-slip transition and the peak y-force disappears. The extraction of the peak y-force is numerically much stabler than the calculation of the stick-slip transition point.

All above observations are in agreement with the assumption that the maximal tangential force applied during the ultrasound bonding defines an upper limit for the shear force given following two conditions: gross ball deformation is absent during ultrasound bonding and intermetallic diffusion can be neglected (bond temperature below approximately 180 °C). The above observations are in the following related to this statement.

The bond growth at the interface is not uniform for ultrasonic bonded wires. Bonded balls that are released from the pad by etching the Al-pad show intermetal-

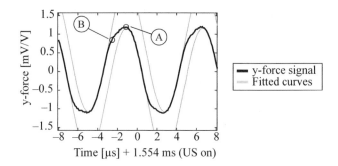

Fig. 5.48. Extraction of the peak y-force (A) and stick-slip transition value (B).

lic islands that grow in number and size with progress in bond growth [29]. To initiate gross sliding, these bonds have to be broken during each ultrasound oscillation. The force that is needed to break these bonds is in first order independent of the applied bond force. If the shear force that is needed for breaking the bonds is higher than the applied tangential force, friction at the contact zone is stopped. In cases where the friction process is the dominating effect for bond growth, the shear force at the end of the bonding is directly related to the tangential force.

The scaling factor (10.5 mV/V/N) between the peak y-force and the shear force is in good accordance with the sensor calibration value (see Sect. 4.2) for measurements performed at 35 °C. For measurements at 163 °C a scaling factor of 7.7 mV/V/N is found. From the sensor calibration a value of 9.28 mV/V/N is expected. A possible explanation for this deviation at elevated temperature is the temperature dependence of the yield stress. The shear test is performed at room temperature whereas the y-force is recorded at 163 °C. Post-baking effects would primarily increase the shear force value for bonds with low adhesion.

The correlation between the peak y-force and the shear force becomes even more obvious, if the time evolution of the shear force compared to the microsensor y-force is displayed (see Fig. 5.49). Each error bar corresponds to average and standard deviation of 10 shear force measurements of balls bonded with the bonding time that corresponds to the error bar position in time. The y-force is extracted from 10 measurement with a total bonding time of 7.5 ms. The peak y-force and the stick-slip transition value are numerically extracted from the y-force. Figure 5.49 shows the mean values (dashed line) and a upper and lower bound (solid line) defined by the standard deviation. The y-force stick-slip transition curve is partly covering the peak y-force signal. The ultrasound current I_{US} is added to enhance clarity. The peak y-force value and the stick-slip transition y-force coincide during the second part of the ultrasound bonding time (3 ms to 7.5 ms). However, the stick-slip transition value is lower during the friction process at the contact zone. The shear force time evolution correlates better with this stick-slip transition value than with the peak y-force.

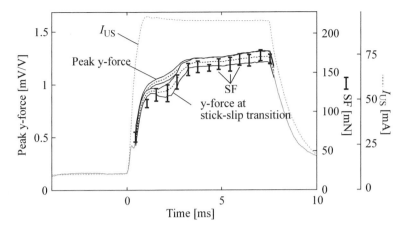

Fig. 5.49. Time evolution of the shear force (SF) compared with the microsensor y-force evaluated at the peak value or at the friction start (stick-slip transition) of the individual oscillations. Each error bar of the SF corresponds to the standard deviation of 10 ball bonds characterized with the shear tester (the balls are bonded with a bonding time corresponding to the error bar positions on the time axis). The peak y-force and the y-force at the stick-slip transition are the mean (dashed line) and standard deviation (solid lines) values of 10 measurements. (F_I = 180 mN, F_N = 87 mN, I_{US} = 91.7 mA, substrate temperature T_s = 163 °C).

The use of microsensor signals as bond process quality indicator is only applicable to production processes if test dies with microsensors were implemented on production wafers. Without sensors on the same wafer material, only equipment drifts could be logged. Therefore, a method that enables in-process measurements is desired. The big advantage of the microsensors is their position close to where bonding happens. The signal to noise ratio will become the critical point for measurement methods that are more remote from the bonding zone [30, 31]. To get full information about the bonding process, two parameters are needed. The first one should give information about the deformation state, the second one should give information about the strength of the bond. Instead of using the microsensor to measure the force in situ and in real-time at ultrasound frequency, it is possible to use the bond head for a quasi static measurement [32].

5.2.6   Comparison Between Au-Al and Au-Au Contacts

In the previous section the characteristics of the Au-Al contact have been described. The pad aluminum is covered by a brittle oxide layer that has to be removed prior to bond formation. This results in a low friction coefficient at the start of the bonding process. For a Au-Au contact no native oxide has to be removed. Thus the adhesion between the gold ball and the gold of the pad is high enough after the impact to

(a)   (b)

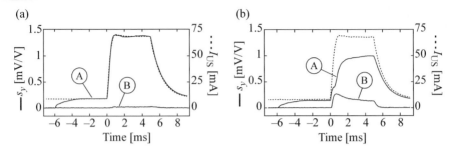

Fig. 5.50. Comparison of the sensor signals of a bond based on a Au-Au contact (**a**) and a Au-Al contact (**b**) for identical bond parameters. The fundamental (A) and the 3rd harmonic (B, scaled by factor 3) of the y-force signal show no indication for friction at the contact zone of a Au-Au contact. The ultrasound amplitude (dashed line) is set below the limit for friction between ball and capillary for these measurements.

inhibit any sliding. Au-Au ball bond measurements are performed on the xyz-sensor *XYZ-Au85*$_{aligned}$ that is covered with a Ni and Au metal layer stack with thickness 5 μm and 1 μm, respectively. Figure 5.50 shows a comparison of the force and machine signals for a Au-Au contact and a Au-Al contact that are bonded with identical settings ($F_I = 200$ mN, $F_N = 100$ mN, $I_{US} = 68$ mA, substrate temperature $T_s = 163$ °C).

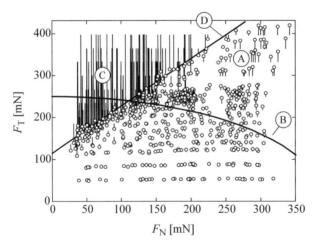

Fig. 5.51. Comparison between forces measured by the microsensor (circles) and the forces calculated from the ultrasound current values (upper end of the vertical lines), joined to circles by vertical lines. Values are identical if no line is visible. The scaling factor used to calculate the "theoretical" tangential force from the ultrasound current is 9.01 mV/V/A. For explanation of A to D see Fig. 5.43. (Sensor sensitivity $g_{yy} = 9.28$ mV/V/N).

The following observations can be made for the Au-Au contact:

- Friction at the contact zone is absent over the whole parameter window.
- Friction between ball and capillary is observed and shows the same characteristics as for Au-Al contacts (see Fig. 5.51). The slope and offset of the demarcating line (D) in Fig. 5.51 are 1.1 and 115 mN, respectively.
- There is no correlation between the peak value of the microsensor tangential force and the off-line measured shear force.
- High shear forces occur already at small ultrasound amplitudes.

5.3   Wedge Bonding Process Knowledge

Compared to the ball bond, gathering knowledge about the second bond (wedge bond) is more complex due to the lower symmetry of the bond geometry and the higher amount of substrate types available. This section concentrates on identification of important processes during a wedge bond using exemplarily selected contact systems Au-Al and Au-Au, i.e. Au wire on Al- or Au-substrate metallizations.

To measure the forces during the wedge bond formation the wedge and ball position is exchanged in the bonder program. The ball is thus bonded on the die pad as shown in Fig. 5.52. The wire direction is of importance for the interpretation of sen-

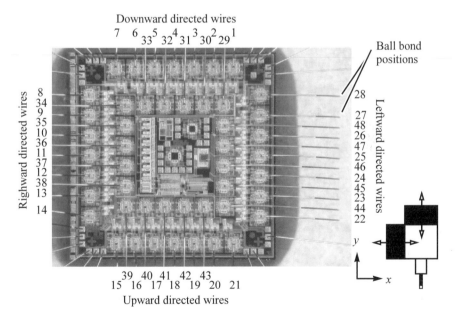

Fig. 5.52. Wedge bond test chip and orientation to the wire bonder. The numbering of the test wire corresponds to the bonding sequence. The ball bond is performed on the die pad and the wedge bond is done on the plated test chip pads.

sor signals due to the reduced symmetry of the wedge bonds. The ball to wedge direction denotes the different wire orientations.

5.3.1   Tail Breaking Force

An important part of the wedge bond is the formation of the tail needed for the next ball. See Sect. 1.2 for detailed information on the wedge bonding sequence. Before the tail is formed by breaking the wire, it must be bonded to the substrate with sufficient strength to withstand the forces during weaving out the wire. The tail bond is formed during the bonding of the wedge. An unstable tail bond results in tail lift-off or tail length variation. A tail lift-off occurs if the tail bond breaks during weaving out the wire from the capillary while the clamp is open. The wire then is completely blown out the capillary and the clamp. This lost wire causes a production stop and the wire has to be threaded again by an operator. Tail length variations is a cause of small balls as a deviant quantity of wire is melt to a ball during electrical flame-off for FAB formation. The force needed for breaking the tail is thus an indicator of the quality of the tail bond and the predetermined breaking point.

The breaking of the tail is observed on the microsensors by thermal and mechanical effects. For heated chip surfaces, the bonding wire in contact with the chip is a heat sink. This results in a temperature gradient on the chip surface close to the bond. As soon as the tail breaks, the conductive path is removed and the temperature on the bonding pad settles at a higher steady state value. Due to the high thermal conductivity of silicon and the close position of the sensors to the bond position, the sensor resistance will change upon a sudden temperature increase. This temperature effect is superimposed on the force that is required to break the wire at the predetermined breaking point for tail formation, as illustrated in Fig. 5.53. After closing the wire clamp, a tensional force is build up by moving the bond head upwards. If the wire breaks the force falls instantly back to zero. The height of this step is equivalent to the force needed to break the wire at its predetermined breaking point. Figure 5.54 shows the signal of the buried z-force sensor (YZZ_b-Au_{Wedge}) during tail formation. The buried force sensor is preferred to the conventional n^+ z-force sensor as its higher sensitivity results in a larger signal to noise ratio. The high-frequency noise on the sensor signal is limiting the accuracy of the tail break-

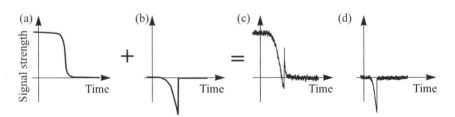

Fig. 5.53. Superposition of cooling effect (**a**) and tail breaking force (**b**) during wedge bond tail breaking at elevated substrate temperatures. Measured sensor signal (**c**) of the tail breaking force of a wedge bonded at substrate temperature $T_s = 172$ °C. Measured sensor signal of the tail breaking force of a wedge bonded at ambient temperature (**d**).

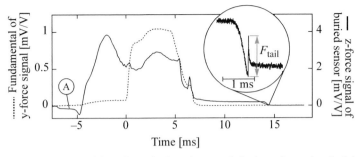

Fig. 5.54. The breaking of the tail results in a sharp peak in the z-force signal of the buried sensor. Heating of the substrate superimposes a signal caused by the cooling of the pad by the wire. The detail view shows the z-force signal during the tail formation. The sensor signal before the impact (A) is explained in the text.

ing force determination. A high sampling rate is needed to separate the tail breaking force from the temperature signal. The superposition of the cooling effect and the tail breaking force is shown in Fig. 5.53. The measured force is negative and decreases during the capillary lift-off process as the wire clamp is pulling the wire. If the wire breaks at the predetermined breaking point the sensor signal relaxes to zero. The rising time of this force relaxation is faster than 2 μs. The time measurement of the rising edge is limited by the employed instrumental amplifier. The measured value for F_{tail} is 174 μV/V or 20.9 mN, correspondingly. The break load of the used gold wires with a diameter of 22 μm is 80 ± 3 mN. This gives an upper limit as a well defined breaking point will have lower strength due to the wire deformation. The tail breaking force strongly depends on the used substrate material, wire and bonding parameters.

The negative force peak (A) prior to the impact in Fig. 5.54 is due to the wire that is touching the reference sensor element. This interfering effect could by reduced in a future sensor design with a larger distance between bond zone and reference sensor element position with the disadvantage of a higher thermal sensitivity.

5.3.2   Gold - Gold Contact Bond

The synthesis of the symmetry properties of the ultrasound oscillation and the wire direction inherently results in a wire direction dependent bonding characteristic of the wedge bond. Figures 5.55 and 5.56 show conventionally measured wedge pull forces and the tail breaking force as function of the impact force, wire direction, and applied ultrasound amplitude. The wedge is bonded on the sensor $YZZ_b\text{-}Au_{Wedge}$ covered with 1 μm Au on 5 μm Ni. The tail breaking force is measured with the buried z-force sensor. Substrate temperature and ultrasound bonding time are 163 °C and 4.6 ms, respectively. The measured bond force is 320 mN.

For small ultrasound amplitudes wedges can only be bonded at high impact forces. At low impact forces, the low deformation results in a poor predetermined

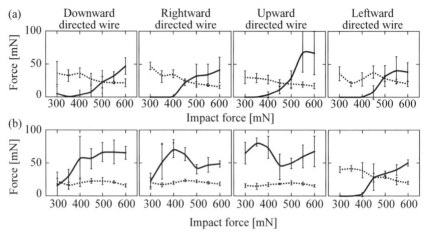

Fig. 5.55. Wedge pull force (solid) and tail breaking force (dashed) for varied impact forces. Tail breaking forces are measured with the buried z-force sensor. Five wedge bonds are recorded for each setting. Measured on a WB 3088iP. $A_H = 49.8$ nm ($I_{US} = 20.4$ mA) (**a**), $A_H = 124.5$ nm ($I_{US} = 51$ mA) (**b**).

breaking point and the wedge is torn from the pad during tail formation. The measured tail breaking force of such wires is exceptionally large. High impact forces result in very strong bonds for upward directed wires for all ultrasound amplitude settings. By shifting to higher ultrasound amplitudes, a peak in the pull values

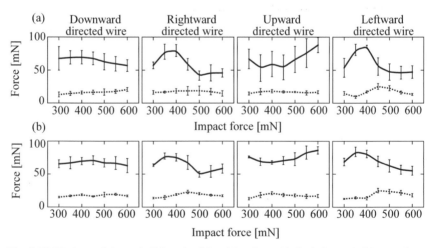

Fig. 5.56. Wedge pull force (solid) and tail breaking force (dashed) for varied impact forces. Tail breaking forces are measured with the buried z-force sensor. Five wedge bonds are recorded for each setting. Measured on a WB 3088iP. $A_H = 248$ nm ($I_{US} = 102$ mA) (**a**), $A_H = 373$ nm ($I_{US} = 153$ mA) (**b**).

appears at impact forces around 350 mN. Ultrasound amplitudes higher than 380 nm (I_{US} = 155 mA) will cause intolerable wire sway, especially for wires bonded with low impact forces.

The measured pull values show significant differences between the four wire directions. Differences in bonding behavior between wires directed perpendicular or in parallel to the ultrasound oscillation are expected. However, the different behavior of upward and downward directed wires can not be explained by considering the ultrasound and wire direction alone. As already seen for ball bonds on an Au substrate, gross friction at the contact zone between pad and wire is absent. As bond formation is no longer dominated by friction processes, wire deformation becomes important for bond formation. The strong functional dependence of the pull force on the applied impact force indicates a high significance of the impact phase on the wedge bond.

Wedge Bond Force Signatures. The ability of the microsensors of measuring forces acting in all three axis directions during the impact and ultrasound bonding phase is used to explore the differences between the wedge bond of various wire to ultrasound oscillation angles. As the ultrasound force oscillation measured on the pad is not symmetrical anymore for positive and negative half waves the envelope with included low frequency forces has to be examined. The inclusion of the low-frequency force signals for bond process analysis is important as the low-frequency tangential forces often rise above the ultrasound induced forces. This is in contrast to ball bonds for which the bonding is dominated by the applied ultrasound forces and normal force for the Au-Al contact system.

In order to measure x-forces during a wedge bond, XYZ-$Au85_{aligned}$ sensors are used in a first step. The advantage of the xyz-force sensor over the specific wedge sensors (YZZ_b-Au_{Wedge}) is its ability of recording simultaneously the forces in all directions. Therefore, forces acting perpendicular to the ultrasound direction can be investigated as well. Small wedges are bonded on Ni-Au coated dies, as the pad metallization is higher than the surrounding passivation. Otherwise, the off-center placed capillary would touch the passivation and the wedge bond process will be distorted. The used wire is of type AW-14 with 20 µm diameter. The capillary DFX-24063-281F-ZP38T from SPT Roth Ltd, Lyss, Switzerland, has a 24 ± 0.5 µm hole diameter, 28 ± 0.5 µm chamfer diameter, and a 11 ° face angle. A wire bonder 3088iP is used for the measurements. Figures 5.57 to 5.60 exemplarily present measured signals of wires with the four main orientations.

Figure 5.57 shows the recorded x- and y-forces of a rightward directed wire. Solid areas in the figures are the measured x-and y-force sensor signals caused by the ultrasound oscillation of the capillary. The individual oscillations are not visible due to the ultrasound frequency of 130 kHz. The machine signal amplitudes of the transducer current, bond force, and z-axis position are plotted for rendering the different bonding phases. They are not to scale. The SEM images on the right side document the side view and top view of the corresponding wedge after the bond process.

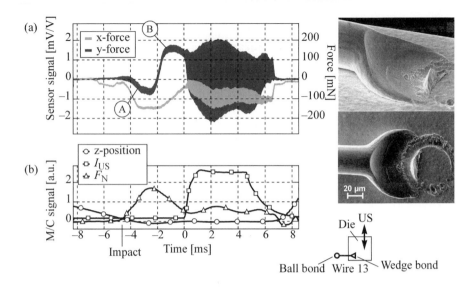

Fig. 5.57. X-force and y-force (**a**) of a rightward directed wire (wire 13). Graph (**b**) shows the machine signals. The impact force to bond force transition results in a y-shift force (A). (F_I = 680 mN, F_N = 240 mN, I_{US} = 83 mA, substrate temperature T_s = 160 °C). The SEM images on the right side document the final state of the wedge bond.

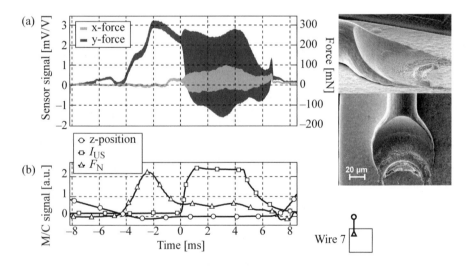

Fig. 5.58. X-force and y-force (**a**) of a downward directed wire (wire 7). The x-force signal exhibits no significant deviation from zero. Graph (**b**) shows the machine signals. (F_I = 880 mN, F_N = 240 mN, I_{US} = 78 mA, substrate temperature T_s = 160 °C). The SEM images on the right side document the final state of the wedge bond.

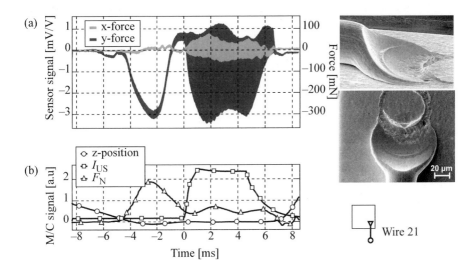

Fig. 5.59. X-force and y-force (**a**) of a upward directed wire (wire 21). Graph (**b**) shows the machine signals. (F_I = 750 mN, F_N = 200 mN, I_{US} = 78 mA, substrate temperature T_s = 160 °C). The SEM images on the right side document the final state of the wedge bond.

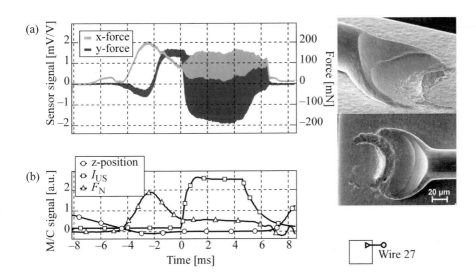

Fig. 5.60. X-force and y-force (**a**) of a leftward directed wire (wire 27). Graph (**b**) shows the machine signals. (F_I = 755 mN, F_N = 220 mN, I_{US} = 83 mA, substrate temperature T_s = 160 °C). The SEM images on the right side document the final state of the wedge bond.

The sketch underneath the micrographs illustrates the wire position on the chip and its direction in comparison to the ultrasound oscillation direction of the capillary tip. As a result of the measured impact force of 680 mN the wire gets already fully deformed during the impact phase as indicated by the constant z-position signal after impact. During the impact phase, both the x-force and the y-force signal show a characteristic time evolution for a specific wire direction. It is important to notice that both signals are strongly deflected from zero. This is in contrast to wires aligned in ultrasound direction as exemplarily shown in Fig. 5.58 for a downward directed wire. The low-frequency signal of the x-force sensor is very small compared to the x-force signal of wires perpendicular to the ultrasound oscillation direction. Similar behavior is found for upward and leftward directed wires as shown in Fig. 5.59 and Fig. 5.60, respectively. The force fingerprint of the leftward directed wire is equivalent to the rightward directed wire with exception of a sign change of the x-force sensor signal.

To verify the reproducibility of these wedge bond force measurements, several wires have been bonded with identical settings. Figure 5.61 shows exemplarily the averaged envelope of recorded y-force signals and the standard deviation of the upper and lower envelope lines. The reproducibility of measurements with identical machine settings and bond directions is high. The deviations are mainly the result of signals due to mechanical vibrations. The following parameters heavily influence the low-frequency behavior of the x- and y-force sensor signals: wire direction, bond dynamics settings, bond material, and bond process parameters such as impact force, bond force, ultrasound amplitude.

Before the ultrasound start, the low-frequency signals can be split up into a component independent from the wire direction and one dependent component. The

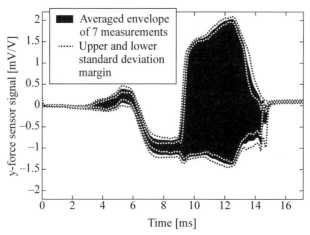

Fig. 5.61. Averaged envelope of the y-force signal of a measurement series of wedge bonds with identical settings. The timing of the individual measurements is relative to the touchdown trigger signal from the wire bonder.

bond direction independent sensor signal component is always associated with the y-direction. Thus, the two signals superpose for wires in y-direction as shown in Fig. 5.62. More precisely, the measured y-force signal of upward or downward directed wires is obtained by subtracting or adding the x-/y-force signals of right-ward directed wires. Measurements on n^+ y-force wedge sensors $YZZ_b\text{-}Au_{Wedge}$ show equivalent low-frequency signatures except for a lower signal strength. The direction independent y-force component is caused by a coupling effect between bond force and tangential force due to the horn bending. For a rightward directed wire as shown in Fig. 5.57, this y-shift force is directed in negative y-direction (A) during impact, whereas the impact to bond force transition results in a tangential force in positive direction (B). Figure 5.63 shows the sensor signal offsets during constant bond force of 200 mN as function of the wire direction. The wire number-ing is defined in Fig. 5.52. All y-force signal offset values are shifted as a result of the wire direction independent y-shift force.

It is observed that the maximum value of the ultrasound force in direction of the ball can reach values that are equivalent to the tangential force of 250 mN to 300 mN during the impact for 20 µm diameter wires. In tail direction, the maximum is much smaller and limited by friction between capillary and the wire as disclosed in the wave form. If these tangential forces reach the plastic limit of the used mate-

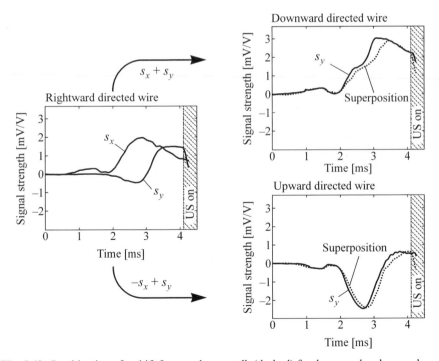

Fig. 5.62. Combination of y-shift force and cross-talk (dashed) for downward and upward directed wires.

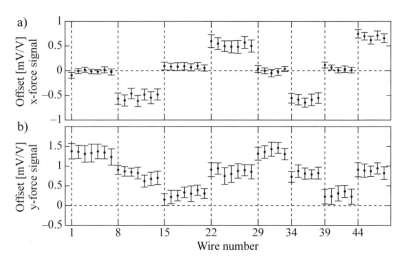

Fig. 5.63. Average and standard deviation of the sensor signal during a constant bond force condition (20 ms duration) between finished impact phase and start of ultrasound. The standard deviation value is dominated by the signal oscillations caused by mechanical vibrations. (F_I = 470 ±15 mN, F_N = 200 ± 10 mN, I_{US} = 78 mA).

rial, plastic deformation will be induced. As the low-frequency force signature depends heavily on the wire direction, high tangential forces are a possible source of direction dependent wedge quality. It is therefore of interest to estimate a typical limit of shear forces that a wedge bond can withstand. This value is extracted with a simple shear test as schematically shown in Fig. 5.64a, measuring the force necessary for shearing the wedge from the pad. The measured force is 103 ± 6 mN for wires with 20 μm diameter bonded at 172 °C on the Ni-Au plated pad. The residuals of the sheared wedge cover an area of 40 × 20 μm^2 as shown by the example in Fig. 5.64b. The measured shear force is to be compared with the forces acting on the wedge during the bond formation.

With the help of additional measurements, the sources of these quite large tangential forces can be identified. Based on measurements with an extended bonding time, low-frequency forces are found on the x-signal that are the result of mechanical vibrations. Their phase and strength depend highly on the machine dynamics settings, and they are especially pronounced for wires bonded in x-direction. Uniform bond quality for all wire directions at high units per hour (UPH) numbers will demand a tight control of such mechanical vibrations (see Sect. 5.1.2).

A further source for a sensor signal that correlates with the wire direction is cross-talk by the applied bond force into the sensor signal of x-/y-force sensors. Due to the lost mirror symmetry plane perpendicular to the wire direction as shown in Fig. 5.65, the symmetry conditions are broken for uniaxial force sensing of the sensor in wire direction (see Sect. 2.1.2). Therefore cross-talk is induced even for a centered contact. The cross-talk properties are characterized in Sect. 4.4 with place-

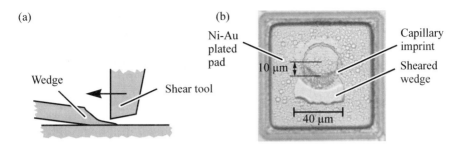

(a)

Wedge

Shear tool

(b)

Ni-Au plated pad

10 μm

40 μm

Capillary imprint

Sheared wedge

Fig. 5.64. Schematic (**a**) of the measurement of the tangential force that is needed to shear the wedge from the pad. The shear height is 2 μm. Micrograph of a sheared wedge (**b**). The capillary position has a 10 μm offset to center the contact zone.

ment sensitivity measurements. Due to the cross-talk any low-frequency force signal of wedge bond measurements has to be carefully interpreted. Based on capillary imprint measurements, an estimation of the cross-talk signal strength under worst-case asymmetric boundary conditions yields a x-force sensor signal offset of 0.36 mV/V for an applied bond force of 200 mN and the capillary off-center placement of 10 μm used for the wedge measurements (see Fig. 5.64b).

Based on an adopted bonding sequence, tangential force relaxations due to plastic flow are observed. The applied normal force during the measurement is 200 mN. No ultrasound is applied during this relaxations phase. The tangential force threshold in wire direction is found at 150 mN, the threshold for forces perpendicular to the wire direction is 100 mN. If ultrasound is turned on, the peak ultrasound amplitude settles at the same levels for large enough ultrasound settings. These force thresholds are higher than the tangential forces measured quasi-statically with the shear tester under absence of any normal force.

The measured x-force signal strength depends on the applied impact force as shown in Fig. 5.66. This functional dependency was only investigated for the x-forces as there exists no additional force component due to the y-shift. For a large ratio between impact force and bond force, the y-force signal indicates force relaxation processes as a result of plastic flow. The z-position signal is recorded before and after the ultrasound bonding phase that is performed after the x-force measure-

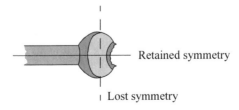

Retained symmetry

Lost symmetry

Fig. 5.65. As the mirror symmetry plane perpendicular to the wire direction is lost for the wedge bond, cross-talk of the applied normal force may be measured in the force sensor in wire direction of a centered wedge.

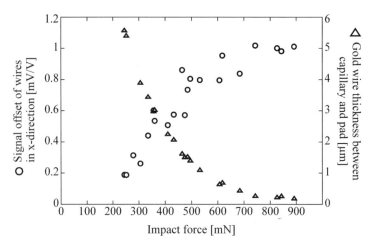

Fig. 5.66. Measured signal offset between finished impact and start of ultrasound of wires in x-direction as function of the applied impact force (see also Fig. 5.63a). An AW-14 wire with 20 μm diameter is used. The capillary DFX-24063-281F-ZP38T has a face angle of 11 °. The bond force is 200 mN. Each measurement point is the average of the 24 x-wires of a test chip. The gold wire thickness indicates the deformation state of the wire during the measurement.

ment. The difference of these two z-positon values is a measure for the wire deformation due to the applied impact force. For impact forces, the wire is hardly deformed. In conclusion, from the measurement on xyz-force sensors following characteristics are observed:

- The low-frequency x- and y-force signals during the impact phase can be interpreted as a superposition of two signals. The first component is identical for all wire directions and found only on the y-force signal. The second component depends on the wire direction.
- The wire direction independent y-force component is caused by the y-shift force due to the applied bond force. This tangential force is dominant for a large ratio between F_I and F_N and is observed during the impact to bond force transition (see Fig. 5.67).
- As a result of the lost symmetry plane perpendicular to the wire direction, a z-force cross-talk impedes the interpretation of the force signal. Based on measurements without bonding wire, i.e. pressing with the capillary directly on the plated test pad, no significant z-force cross-talk is found on the x-force signal for off-center placed bond positions.
- The deformation of the wire can result in a force directed towards the ball due to the wedge-shaped capillary tip pushing against the wire.
- For wires bonded in x-direction on a wire bonder 3088iP, a low-frequency oscillation is superimposed on the x-force sensor signal. The source of these oscilla-

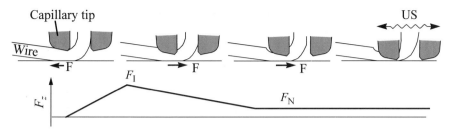

Fig. 5.67. Schematic drawing of the different deformation stages during wedge bonding. The tangential force that is applied on the pad during the different phases is indicated with an arrow.

tions with a frequency of ~ 60 Hz are mechanical vibrations of the bond head. The strength of this vibration component strongly depends on the employed speed during the loop forming and wedge search process and can be reduced to sufficiently low levels.

5.3.3 Gold - Aluminum Contact Bond

A Au-Al wedge bond shows a similar behavior as the Au-ball on a Al-pad, where friction processes dominate. Figure 5.68a shows the time evolution of the off-line measured pull force of wedges bonded with identical settings except for varied bonding times. The different bonding phases can be identified using the fundamen-

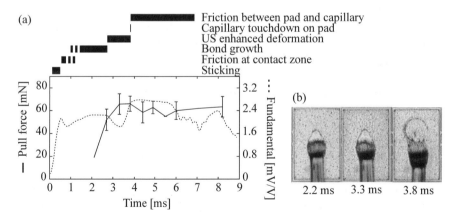

Fig. 5.68. Time evolution (**a**) of the pull force value of the Au-Al wedge bond in comparison with the measured ultrasound y-force signal (dashed line). A special buried y-force microsensor is used for the measurement [33]. Micrographs of the wedge for different bonding times (**b**). ($F_I = 400$ mN, $F_N = 250$ mN, $A_H = 350$ nm ($I_{US} = 143.5$ mA), substrate temperature $T_s = 163$ °C, 22 μm wire diameter, capillary SBNE-28ZA-AZM-1/16-XL-50MTA).

tal of the y-force microsensor signal. Further measurements and the used experimental setup are described in [33].

The following results are found:

- The y-force microsensor signal at the start of the ultrasound enhanced deformation is predominantly independent of applied ultrasound power and temperature. Ultrasound current amplitude and bond temperature were varied from 70 mA to 120 mA and from 40 °C to 160 °C, respectively.
- The start of ultrasound enhanced deformation is tightly connected with the friction duration and depends strongly on the bonding parameters bond force and ultrasound amplitude.
- Process parameter settings with a large impact force result in a poor pull force value for substrate temperatures below 200 °C. Furthermore, tail lift-off problems exist as a result of a too weak tail bond.

Wedge Bond Force Signatures. Equivalent to Ni-Au plated pads, the force signatures are recorded with xyz-force sensors on the Al-Au contact system. Figures 5.69 to 5.71 exemplarily show force signals for the different wire directions. An AW-14 wire with 20 μm diameter and a DFX-24063-281F-ZP38T capillary are used for the measurements.

As the Al test pad is lower than the surrounding passivation of the chip, only small wires can be bonded on these sensors. Otherwise the capillary would touch the passivation. The x- and y-force variations during the impact phase are caused by the wire that is touching the passivation at the border of the pad opening.

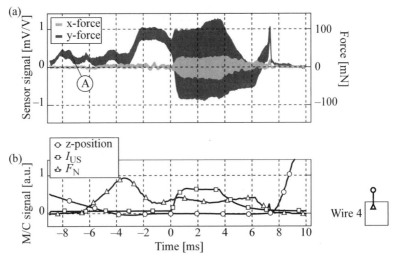

Fig. 5.69. X-force and y-force (**a**) of an upward directed wire bonded on an Al-pad (wire 4). Graph (**b**) shows the machine signals. (F_I = 370 mN, F_N = 140 mN, I_{US} = 55.5 mA, substrate temperature T_s = 160 °C).

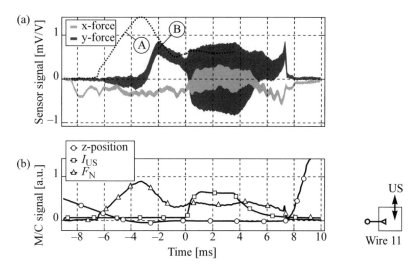

Fig. 5.70. X-force and y-force (**a**) of a leftward directed wire bonded on an Al-pad (wire 11). Graph (**b**) shows the machine signals. (F_I = 355 mN, F_N = 135 mN, I_{US} = 55.5 mA, substrate temperature T_s = 160 ºC).

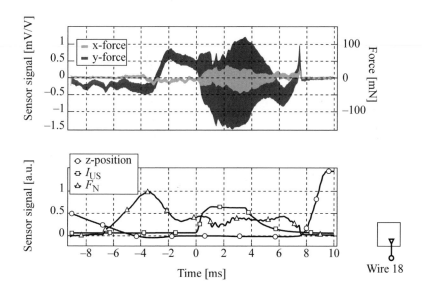

Fig. 5.71. X-force and y-force (**a**) of a downward directed wire bonded on an Al-pad (wire 18). Graph (**b**) shows the machine signals. (F_I = 370 mN, F_N = 140 mN, I_{US} = 55.5 mA, substrate temperature T_s = 160 ºC).

Due to its close position to the sensor structures, a signal is measured. This signal fluctuations on the xyz-force ball bond sensor are always seen on the sensor that is aligned in wire direction as e.g. at time (A) in Fig. 5.69. As the dedicated wedge sensors have larger pad sizes in wire direction, this effect is not observed anymore.

Similar to the Al-Au ball contact, friction at the contact zone plays an important role during the wedge bond formation on an Al substrate as well. The maximum force that can be built up is limited by friction effects and is therefore a function of the applied normal force. The dashed line (A) in Fig. 5.70 stands for the friction limit of the tangential force calculated from the bond force. Due to the bond force decrease, the y-shift force increases until point (B). From this point onwards, the maximum tangential force decreases even though further y-shift should occur. This decrease is to be explained by friction processes.

5.4   Flip-Chip Application

Due to the inaccessibility of the solder joints, flip-chip process optimization is mainly based on FEM simulations and time consuming temperature cycling tests. The presented test chip monitors in real-time and in situ the forces at the interconnection joints, promising to be an efficient tool for flip-chip process development and optimization, as well as examinations of material and reliability issues. Complementing other packaging test chips for flip-chip characterization [34, 35], this sensor measures forces directly at the solder joints [36]. In contrast to other experimental flip-chip measurement setups [37], the integrated force sensors allow the determination of the force that is acting on the individual solder balls. The ability to measure forces with frequencies up to 1 MHz furthermore allows for combined vibrational and thermal stressing for packaging reliability investigations.

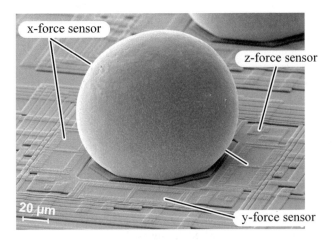

Fig. 5.72. Micrograph of a solder ball with surrounding 3-D force sensor.

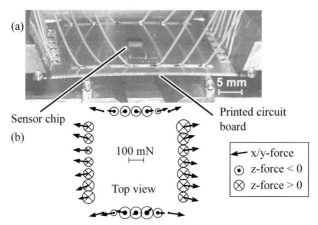

Fig. 5.73. Micrograph of flip-chip test die mounted on a four point bending bridge (**a**) and measured force vectors of solder balls (**b**).

A Ni-Au under bump metallization (UBM) is used as a base for the solder balls on the flip-chip test die, as shown in Fig. 5.72. The solder joints under measurement serve also the purpose of electrical interconnection for the read-out of the sensor signals. For all flip-chip experiments, only the outer sensors were supplied with solder balls. The inner sensors were therefore not in mechanical contact with the substrate, as no underfill was used. Sn63Pb37 solder balls with a diameter of 105 μm were bumped on the Ni-Au UBM by Pac Tech, Nauen, Germany. The pad for the solder balls is octagonal with a minimum pad opening diameter of 85 μm. No underfill was applied, as underfill touching the surface of the sensors would change their sensitivity. The substrates were made from a high temperature FR4 (0.76 mm thickness, $T_g = 180\ °C$) with a Ni-Au finish on 12 μm thick Cu lines. Figure 5.73 illustrates the setup and results of simultaneous measurement of the force vectors acting on the 28 solder balls during bending on a four point bending bridge. Under excessive bending, force saturation caused by plastic flow in the solder is observed on the balls on the left and right side. The in situ and real-time recording of the force signals can be performed until the first contact fails.

5.4.1   Thermal Cycling

During thermal cycling, the elasto-plastic property of the solder ball determines the shape of the sensor signal curves. If the stress in the solder ball is below the plastic limit during the cycles, the measured force depends on the history. Prior loading results in a long term drift of the measured force. However, as soon as plastic flow occurs, the shape of the force profile becomes stationary after the first cycle. Temperature gradients in the substrate can intensify the measured forces. Thus, the temperature has to be tightly controlled during cycling.

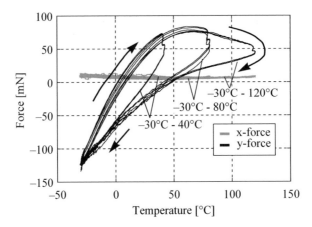

Fig. 5.74. X-force and y-force signals of pad 18 as a function of temperature during temperature cycling. Each hysteresis loop consists of 4 cycles. The x-force signal is constant as this sensor lies on the x-axis of symmetry. Force zero point at start of cycling. Preloading due to the cooling down after the flip-chip bonding results in an initial offset.

Figure 5.74 shows the x-force and y-force hysteresis loops during thermal cycling between different temperature ranges. The cycle time is 150 min (60 min ramp up, 90 min ramp down) for all cycles. The ramp time includes at minimum a 10 min soak time. Only stationary force loops are shown. Dents in the force curves are the result of temperature oscillations around the set point. The force zero point in Fig. 5.74 is at start of cycling. The cooling down after the flip-chip bonding results in a prestressing of the device. The time history of the force during the cycling is plotted in Fig. 5.75. For upper cycling temperatures of 120 °C, the entire force relaxation is observed during the soak time. Similar measurements were performed on soldered surface mount ceramic chip carriers [38]. A FEM model for this configuration is presented in [39]. The yield stress σ_{yield} of Sn63Pb37 solder is a function of the temperature

$$\sigma_{yield}(T) = \alpha_0 + \alpha_1 T + O(T^2),\qquad(5.23)$$

whereas T is expressed in units of °C and the constants are $\alpha_0 = 34.43$ MPa and $\alpha_1 = -0.306$ MPa K^{-1} for Sn63Pb37 solder [39]. The plastic force relaxation for high temperatures in Fig. 5.75 is a result of the temperature dependence of the yield stress described by equation (5.23). The force at the yield point is calculated with relation (5.22) and the contact zone diameter.

For small I/O counts thermosonic flip-chip bonding is an alternative to the flip-chip bonding based on solder balls. Plated or thermosonic bonded Au bumps form the contacts between die and substrate. For thermosonic bonding of the flip-chip, the substrate to die alignment is crucial to achieve a uniform distribution of the

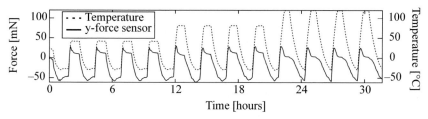

Fig. 5.75. Time history plot of the y-force of sensor 18 during thermal cycling. The entire force relaxation occurs during the soak time for temperatures of 120 °C.

ultrasound energy to the individual pads [40]. The xyz-force sensor ability to measure ultrasound forces offers in situ and real-time monitoring of the forces acting on the individual contacts.

6 Conclusions and Outlook

Conclusions

A novel triaxial force sensor has been developed for measuring the forces acting on the electro-mechanical contacts of wire bonding and flip-chip packaging processes. The sensing principle is based on the piezoresistivity of the source/drain implantation of a commercially available 0.8 μm, double metal, single polysilicon CMOS process. The sensor design makes use of the symmetry properties of the Wheatstone bridge and the stress field to achieve uniaxial force sensing. The sensitivity and cross-talk between the different axes has been described by a tensor formalism.

These xyz-force sensors are the core of a packaging test chip family for wire bonding and flip-chip process investigations. Up to 48 xyz-force sensors connected to a multiplexer bus enable the recording of large measurement series with minimal set-up times. The packaging test chip is embedded in a real-time, high speed measurement system that is able to record up to 20 measurement recurrences (wires) per second. During each measurement the complete set of microsensor signals and wire bonding reference signals are recorded. The large frequency range of the sensor allows for the recording of low-frequency signals, as e.g. machine vibrations, as well as the individual oscillations of the ultrasound vibrations during the thermosonic bonding process. This in situ, real-time inspection method is used for wire bond machine and process research and development.

Based on the developed force sensors, machine parameters including ultrasound amplitude and applied normal force have been related to physical quantities at the bonding zone. Until now, the inaccessibility to the bonding zone prevented inspection of the process window in terms of physical quantities acting at the bond zone itself. Furthermore, off-line inspection methods only qualified the final stage of bonding. The xyz-force sensor gives the opportunity to resolve the time evolution of the bond process. The bond process can be divided into different stages according to the various dominating processes. Physical processes, as e.g. friction at contact zone, friction between ball and capillary, have been identified by changes in the ultrasound wave form of the microsensor signal. The friction between ball and capillary limits the maximum attainable shear force for small bond forces. In addition, the starting point of ultrasound enhanced deformation has been described by an elliptical line in the parameter space of the normal and tangential force. Over the whole process window, the maximal tangential force measured by the microsensors correlates well with the off-line measured shear force for bonds using the Au-Al contact system.

In contrast to the Au-Al contact system, friction at the contact zone is absent for both the first (ball) and the second bond (wedge) using Au-Au contacts. The force that is needed for tail formation can be measured with the z-force sensor. An explanation of the direction dependence of the bond strength during the second bond are low-frequency forces that are present during the deformation process.

The xyz-force sensor has also been used to measure the force acting on the individual solder balls of a flip-chip during thermal cycling and substrate bending.

Outlook

To further develop the wire bonding process towards finer pitches, more reliable contacts, higher yields, and lower cost, an improved understanding of the physics of the process will be helpful. To assess the quality of the required physical process models, the experimental knowledge of the physical quantities at the bonding zone is a prerequisite. The force measurements of the microsensor are thus a good starting point for modelling the bond growth mechanism. The bonding parameter dependence of the offset and slope of the demarcating line of friction between ball / capillary in the ultrasound / normal force parameter plane has to be examined further. The bond parameters close to that line are a promising starting point for bond process optimization focused on decreasing process sensitivity to ultrasound variation (e.g. caused by a capillary exchange). The point of coinciding lines of deformation and friction between ball / capillary is of peculiar interest as it offers small deformation in combination with high shear force and process stability.

This work concentrated on bond growth effects that take place at low substrate temperatures, e.g. during bonding of BGA-substrates. At high substrate temperatures, intermetallic phase growth due to diffusion processes influences the bond strength. To extend a friction based bond growth model to higher temperatures, thermal effects need to be included. It has been shown that gross friction processes at the contact zone are absent for Au-Au contact systems. Further process investigations are needed to understand the influence of the ultrasound on the bond strength.

For future bonding processes, the tangential forces acting on the wedge bond during the impact and ultrasound bonding time could be controlled with appropriate bond head movement, resulting in a further reduction in the direction dependence. This work concentrated on the investigation of Au-Al and Au-Au contact systems. The deposition of a Ag metallization would extend the selection of important contact systems for wedge bond investigations. Further optimizations of the wedge bond sensors should concentrate on reduction of detrimental cross-coupling of the bond force in the x- and y-forces.

The sensor system is optimized for the wire bonding application. Thus a design with decreased sensitivity to thermal stresses in the CMOS layers could significantly improve the temperature range for low-frequency force measurement as is needed for the flip-chip applications.

Abbreviations, Symbols, and Definitions

Abbreviations

ADC	Analog to digital converter
BDC	Ball diameter at capillary imprint
BGA	Ball grid array
BH	Ball height
CERDIP	Ceramic dual-in-line package
CONT	Contact (removal of contact oxide)
CMOS	Complementary metal oxide semiconductor
DAC	Digital to analog converter
DIFF	Diffusion
EFO	Electrical flame-off
FAB	Free air ball (gold ball before impact)
FC	Flip-chip
FEM	Finite element method
FFT	Fast Fourier transform
FIR	Finite impulse response
FOX	Field oxide
FR4	Epoxy glass fibre laminate
FPGA	Field Programmable Gate Array
FSR	Full scale range
H	Capillary hole diameter
IIR	Infinite impulse response
KOH	Potassium hydroxide
LED	Light emitting diode
LIM	Laser vibrometer measurement
LSL	Lower specification limit
MEMS	Micro electro mechanical system
MOSFET	Metal oxide semiconductor field effect transistor
NEF	Noise equivalent force
PCB	Printed circuit board
PE PSG	Plasma enhanced phosphorus silica glass
PSD	Position sensitive detector

PTAT	Proportional to absolute temperature	
QFN	Quad-flat non-leaded	
QFP	Quad-flat package	
RMS	Root mean square	
SEM	Scanning electron microscope	
S/N ratio	Signal to noise ratio	
SF	Shear force	
SS	Shear strength (SF normalized by ball contact area)	
SOI	Silicon on insulator	
TC	Temperature coefficient	
TCE	Thermal expansion coefficient	
TCO	Temperature coefficient of sensor offset	
TCS	Temperature coefficient of sensor sensitivity	
TCR	Temperature coefficient of resistivity	
T_g	Glass transition temperature	
UBM	Under bump metallization	
UPH	Units per hour	
US	Ultrasound	
USL	Upper specification limit	
VDD	Positive supply voltage	
VSS	Negative supply voltage	

Symbols

A_H	Ultrasound amplitude at horn	[m]
A_T	Ultrasound amplitude at freely vibrating capillary tip	[m]
E	E-modulus	[Pa]
Φ	Electrostatic potential	[V]
F_j	Force applied in direction j	[N]
F_N	Bond force measured by m/c sensor	[N]
F_I	Impact force measured by m/c sensor	[N]
g_{ij}	Transductance tensor	[V/V/N]
I	Cross-sectional-area moment of inertia	[m^4]
I_{US}	Ultrasound current amplitude	[A]
κ	Timoshenko shear coefficient	[ ]
Λ_{ij}	Conductivity tensor	[A V^{-1}m^{-1}]
μ	Dynamic friction coefficient	[ ]
μ_0	Static friction coefficient	[ ]
ν	Poison-ratio	[ ]
π_{ijkl}	Piezoresistive tensor	[Pa^{-1}]
r_c	Contact radius	[m]

S_i	Sensor signal of sensor i	[V]
s_i	Normalized sensor signal of sensor i	[V/V]
$\tilde{s}_i$	s_i of linearized Wheatstone bridge	[V/V]
σ_{ij}	Stress field tensor	[Pa]
σ_{yield}	Yield stress	[Pa]
T_s	Substrate temperature	[°C]
t_{US}	Ultrasound bonding time	[ms]
U	Voltage across the Wheatstone bridge	[V]
E_{cap}	E-modulus of capillary	[Pa]
c_{cap}	Viscous damping constant of capillary	[kg m^{-1}s^{-1}]
ρ_{cap}	Density of capillary	[kg m^{-3}]
ν_{cap}	Poisson-ratio of capillary	[ ]
K_{TH}	Rotational spring constant transducer horn	[N m rad^{-1}]
c_{cz}	Damping constant contact zone	[kg s^{-1}]
K_{Tcz}	Rotational spring constant contact zone	[N m rad^{-1}]
K_{cz}	Spring constant contact zone	[N m^{-1}]
m_{cz}	Lumped mass of contact zone	[kg]
c_{Tcz}	Rotational damping constant contact zone	[N m s rad^{-1}]

Definitions

$$\partial_z^2 \equiv \frac{\partial^2}{\partial z^2}$$ Partial derivative

$$i \equiv \sqrt{-1}$$

References

Chapter 1

1. M. Mayer, "Microelectronic Bonding Process Monitoring by Integrated Sensors," Ph.D. Thesis, No. 13685, ETH Zurich, Zurich, 2000.
2. R. van Gestel, "Reliability Related Research on Plastic IC-Packages: A Test Chip Approach," Ph.D. Thesis, Delft University Press, 1993.
3. J.N. Sweet, "Integrated Test Chips Improve IC Assembly" IEEE Circuits and Devices Magazine, 6, pp. 39–45, 1990.
4. S. J. Ham, S. B. Lee, "Measurement of Creep and Relaxation Behaviors of Wafer-Level CSP Assembly Using Moiré Interferometry," Transactions of the ASME, 125, pp. 282–288, 2003.
5. D. A. Bittle, J. C. Suhling, R. E. Beaty, R. C. Jaeger, R. W. Johnson, "Piezoresistive Stress Sensors for Structural Analysis of Electronic Packages," Journal of Electronic Packaging, 113, pp. 203–215, 1991.
6. J. F. Creemer, "The Effect of Mechanical Stress on Bipolar Transistor Characteristics," Ph.D. Thesis, Techn. Univ. Delft, Delft, 2002.
7. A. T. Bradley, R. C. Jaeger, J. C. Suhling, K. J. O'Connor, "Piezoresistive Characteristics of Short-Channel MOSFETs on (100) Silicon," IEEE Trans. on Electron Devices, 48, No. 9, pp. 2009–2015, 2001.
8. M. Doelle, P. Ruther, O. Paul, "A Novel Stress Sensor Based on the Transverse Pseudo-Hall Effect of MOSFETS," IEEE 16th Annual International Conference on MEMS, pp. 490–493, 2003.
9. Y. Zou, J. C. Suhling, R. C. Jaeger, S. T. Lin, J. T. Benoit, R. R. Grzybowski, "Die Surface Stress Variation During Thermal Cycling and Thermal Aging Reliability Tests," 49th Electronic Components and Technology Conference, pp. 1249–1260, 1999.
10. Y. Zou, J. C. Suhling, R. W. Johnson, R. C. Jaeger, A. K. M. Mian, "In-situ Stress State Measurements During Chip-on-Board Assembly," IEEE Transactions on Electronics Packaging Manufacturing, 22, No. 1, pp. 38–52, 1999.
11. P. Palaniappan, D. F. Baldwin, "In Process Stress Analysis of Flip-Chip Assemblies During Underfill Cure," Microelectronics Reliability, 40, pp. 1181–1190, 2000.

12.   J. Zhang, H. Ding, D. F. Baldwin, I. C. Ume, "Characterization of In-Process Substrate Warpage of Underfilled Flip Chip Assembly," IEEE/CPMT/SEMI 28th Int'l Electronics Manufacturing Technology Symposium, pp. 291–297, 2003.

13.   M. Mayer, J. Schwizer, O. Paul, H. Baltes, "In-situ Ultrasonic Stress Measurement During Ball Bonding using Integrated Piezoresistive Microsensors," Proc. Intersociety Electron. Pack. Conf. (InterPACK99), pp. 973–978, 1999.

14.   D. Manic, A. P. Friedrich, Y. Haddab, R. S. Popovic, "Influence of Assembly Procedure on IC Parameters," Proc. 21st International Conference on Microelectronics, **2**, pp. 637–640, 1997.

15.   M. Doelle, C. Peters, P. Gieschke, P. Ruther, O. Paul, "Two-Dimensional High Density Piezo-FET Stress Sensor Arrays for In-Situ Monitoring of Wire Bonding Processes," Proc. 17th IEEE International Conference on Micro Electro Mechanical Systems (MEMS'04), pp. 829–832, 2004.

16.   M. Hizukuri, Y. Wada, N. Watanabe, T. Asano, "Real Time Measurement of the Strain Generated on Substrate during Ultrasonic Flip Chip Bonding," 6th Symposium on Microjoining and Assembly Technology in Electronics, 3-4 February, Yokohama, pp. 169–174, 2000.

17.   M. Hizukuri, T. Asano, "Measurement of Dynamic Strain During Ultrasonic Au Bump Formation on Si Chip," Japanese Journal of Applied Physics, Part 1 (Regular Papers, Short Notes & Review Papers), **39**, No. 4B, pp. 2478–2482, 2000.

18.   M. Mayer, Z. Stoessel, D. Bolliger, O. Paul, "Process and Chip for Calibrating a Wire Bonder," EP patent application publication No. 0953398A1, November 3, 1999.

19.   G. Harman, "Wire Bonding in Microelectronics," 2nd Ed., McGraw Hill, New York, 1997.

20.   M. Barp, D. Vischer, "Achieving a World Record in Ultra High Speed Wire Bonding through Novel Technology," Proc. Semicon West'02, 2002.

21.   N. Onda, Z. Stössel, A. Dommann, J. Ramm, "DC-Hydrogen Plasma Cleaning: A Novel Process for IC-Packaging," Proc. Semicon / Package Materials, pp. D1–D10, 1997.

22.   D. S. Liu, Y. C. Chao, C. H. Wang, "Study of Wire Bonding Looping Formation in the Electronic Packaging Process Using the Three-Dimensional Finite Element Method," Finite Elements in Alanlysis and Design, **40**, pp. 263–286, 2004.

23.   ASTM Standard No. F1269–89, "Test Methods of Destructive Shear Testing of Ball Bonds," 1990.

24.   F. Farassat, "Wire Bonding Ultrasonic Control System Responsive to Wire Deformation," US patent application publication No. 5314105, May 24, 1994.

25. F. Farassat, "Entwicklung und Erprobung eines Regelsystems zur Verbesserung der Verbindungsqualität beim Ultraschallbonden," (in German), Ph.D. Thesis, Technical University Berlin, Berlin, 1997.

26. A. Carrass, V. P. Jaecklin, "Analytical Methods to Characterize the Interconnection Quality of Gold Ball Bonds," 2nd European Conference on Electronic Packaging Technology (EuPac '96), DVS Berichte, **173**, pp. 135–139, 1996.

27. R. Pufall, "Automatic Process Control of Wire Bonding," Proc. Electronic Components & Technology Conference, **43**, pp. 159–162, 1993.

28. J. H. Cusick, A. E. Brown, A. S. Hamamoto, J. L. S. Bellin, "Ultrasonic Bond Monitor," US patent application publication No. 3890831, June 24, 1975.

29. K. U. von Raben, "Controlling Relevant Bonding Parameters of Modern Bonders," Proc. Electronic Components Conference, **38**, pp. 558–563, 1988.

30. S. W. Or, H. L. W. Chan, V. C. Lo, C. W. Yuen, "Ultrasonic Wire-Bond Quality Monitoring Using Piezoelectric Sensor," Sensors and Actuators, **A65**, pp. 69–75, 1998.

31. W. L. Loofbourrow, "Capacitive Microphone Tuning of Ultrasonic/Thermosonic Bonders," Proc. Semiconductor Processing, pp. 472–484, 1984.

32. K. D. Lang, F. Osterwald, B. Schilde, H. Reichl, "Measurement of Ultrasonic Behavior During Wire Bonding - a Contribution to Quality Assurance in Chip on Board Technology," Proc. Semicon West '98, pp. F1–F9, 1998.

33. A. Schneuwly, P. Gröning, L. Schlapbach, G. Müller, "Bondability Analysis of Bond Pads by Thermoelectric Temperature Measurements," Journal of Electronic Materials, **27**, No. 11, pp. 1254–1261, 1998.

34. M. Mayer, O. Paul, D. Bolliger, H. Baltes, "Integrated Temperature Microsensors for Characterization and Optimization of Thermosonic Ball Bonding Process," Proc. Electronic Components and Technology Conference, pp. 463–468, 1999.

35. O. E. Gibson, "Bond Signature Analyzer," US patent application publication 4998664, March 12, 1991.

36. J. Chen, "Real-Time Ultrasonic Testing of Quality of Wire Bonding," Insight Non-Destructive Testing and Condition Monitoring, **44**, No. 7, pp. 443–445, 2002.

Chapter 2

1. J. F. Nye, "Physical Properties of Crystals," Oxford University Press, Oxford, 1957.

2. T. Toriyama, S. Sugiyama, "Analysis of Piezoresistance in p-Type Silicon for Mechanical Sensors," Journal of Microelectromechanical Systems, **11**, No. 5, pp. 598–604, 2002.

3. J. C. Suhling, R. C. Jaeger, "Silicon Piezoresistive Stress Sensors and Their Application in Electronic Packaging," IEEE Sensors Journal, **1**, No. 1, pp. 14–29, 2001.

4.   A. Nathan, H. Baltes, "Microtransducer CAD, Physical and Computional Aspects," Springer, Vienna, 1999.
5.   —, "X-CMOS 0.8, Modular Mixed Signal Technology," http://www.xfab.com/sheets/ds-cx08.pdf, Datasheet, X-FAB Semiconductor Foundries AG, 2002.
6.   O. N. Tufte, E. L. Stelzer, "Piezoresistive Properties of Silicon Diffused Layers," Journal of Applied Physics, **34**, No. 2, pp. 313–317, 1963.
7.   Y. Kanda, "A Graphical Representation of the Piezoresistance Coefficients in Silicon," IEEE Transactions on Electron Devices, **29**, No. 1, pp. 64–70, 1982.
8.   U. Schiller, "Thermomechanical Offset in Integrated Hall Plates," Diploma Thesis, IMTEK, University of Freiburg, Freiburg, 2001.
9.   B. L. Lwo, C. H. Kao, T. S. Chen, Y. S. Chen, "On the Study of Piezoresistive Stress Sensors for Microelectronic Packaging," Journal of Electronic Packaging, **124**, pp. 22–26, 2002.
10.   B. J. Lwo, T. S. Chen, C. H. Kao, Y. L. Lin, "In-Plane Packaging Stress Measurement Through Piezoresistive Sensors," Journal of Electronic Packaging, **124**, pp. 115–121, 2002.
11.   W. Pietrenko, "Einfluss von Temperatur und Störstellenkonzentration auf den Piezowiderstandseffekt in n-Silizium," (in German), Physica Status Solidi, A, No. 41, pp. 197–205, 1977.
12.   S. F. Chu, "Piezoresistive Properties of Boron and Phosphorous Implanted Layers in Silicon," Ph.D. Thesis, Case Western Reserve University, 1978.
13.   M. Mayer, "Microelectronic Bonding Process Monitoring by Integrated Sensors," Ph.D. Thesis, No. 13685, ETH Zurich, Zurich, 2000.
14.   S. A. Liu, H. L. Tzo, "A novel six-component force sensor of good measurement isotropy and sensitivities," Sensors and Actuators A, **100**, pp. 223–230, 2002.
15.   V. I. Fabrikant, "Applications of Potential Theory in Mechanics," Kluwer Academic Publishers, Dordrecht, 1989.
16.   J. Rosen, "Symmetry in Science," Springer, New York, 1995.
17.   M. Hizukuri, T. Asano, "Measurement of Dynamic Strain During Ultrasonic Au Bump Formation on Si Chip," Japanese Journal of Applied Physics, Part 1 (Regular Papers, Short Notes & Review Papers), **39**, No. 4B, pp. 2478–2482, 2000.
18.   T. Ikeda, N. Miyazaki, K. Kudo, K. Arita, H. Yakiyama, "Failure Estimation of Semiconductor Chip During Wire Bonding Process," Journal of Electronic Packaging, **121**, pp. 85–91, 1999.
19.   A. C. Fischer-Cripps, "Introduction to Contact Mechanics," Springer, New York, 2000.
20.   Y. Takahashi, S. Shibamoto, K. Inoue, "Numerical Analysis of the Interfacial Contact Process in Wire Thermocompression Bonding," IEEE Transactions on Components, Packaging, and Manufacturing Technology, Part A, **19**, No. 2, pp. 213–223, 1996.

21. I. C. Noyan, "Plastic Deformation of Solid Spheres," Philosophical Magazine A, **57**, No. 1, pp. 127–141, 1988.

22. W. Budweiser, "Untersuchung des Thermosonic Ballbondverfahrens," (in German), Ph.D. Thesis, Technical Univ. Berlin, Berlin, 1993.

23. J. Schwizer, M. Mayer, D. Bolliger, O. Paul, H. Baltes, "Thermosonic Ball Bonding: Friction Model Based on Integrated Microsensor Measurements," Proc. 25th IEEE/CPMT Intl. Electronics Manufacturing Technology Symposium IEMT, pp. 108–114, 1999.

24. K. L. Johnson, "Contact Mechanics," Cambridge University Press, Cambridge, pp. 70–74, 1985.

25. L. D. Landau, E. M. Lifschitz, "Theory of Elasticity," Pergamon Press, London, 1959.

26. J. R. Barber, "Elasticity," Kluwer Academic Publishers, Dordrecht, 1992.

27. G. M. Hamilton, L. E. Goodman, "The Stress Field Created by a Circular Sliding Contact," Trans. ASME, Journal of Applied Mechanics, **33**, pp. 371–376, 1966.

28. O. Madelung, M. Schulz, H. Weiss, "Landolt-Börnstein," Semiconductors, **17**, Berlin, 1982.

29. A. Schroth, "Modelle für Balken und Platten in der Mikromechanik," Dresden University Press, Dresden, 1996.

30. V. Ziebart, "Mechanical Properties of CMOS Thin Films," Ph.D. Thesis, No. 13457, ETH Zurich, Zurich, 1999.

31. J. H. Lau, "Thermal Stress and Strain in Microelectronics Packaging," Van Nostrand Reinhold, New York, 1993.

32. D. S. Gardner, P. A. Flinn, "Mechanical Stress as a Function of Temperature in Aluminum Films," IEEE Transactions on Electron Devices, **35**, No. 12, pp. 2160–2169, 1988.

33. A. Carrass, V. P. Jaecklin, "Analytical Methods to Characterise the Interconnection Quality of Gold Ball Bonds," Proc. 2nd Europ. Conf. Electr. Packaging Technol. EuPac '96, pp. 135–139, 1996.

34. M. L. Minges, "Packaging," Electronic Materials Handbook, **1**, ASM International, Materials Park, pp. 534ff, 1989.

35. M. Mayer, J. Schwizer, O. Paul, H. Baltes, "In-situ Ultrasonic Stress Measurement During Ball Bonding using Integrated Piezoresistive Microsensors," Proc. Intersociety Electron. Pack. Conf. (InterPACK99), pp. 973–978, 1999.

36. J. Schwizer, M. Mayer, O. Brand, H. Baltes, "Analysis of Ultrasonic Wire Bonding by In situ Piezoresistive Microsensors," Proc. Transducers '01 / Eurosensors XV, pp. 1426–1429, 2001.

37. J. Schwizer, M. Mayer, O. Brand, H. Baltes, "In Situ Ultrasonic Stress Microsensor for Second Bond Characterization," Proc. International Symposium on Microelectronics IMAPS, pp. 338–343, 2001.

Chapter 3

1. J. Schwizer, "In-situ Wire-Bond Prozessuntersuchung mit integrierten piezoresistiven Mikrosensoren" (in German), Diploma Thesis, Physical Electronics Laboratory, ETH Zurich, Zurich, 1999.
2. D. Gabor, "Theory of Communication," Journal of the Institution of Electrical Engineers, **93**, pp. 429–457, 1946.
3. B. Jähne, "Digital Image Processing," Springer, 5th Edition, 2002.
4. M. Cerna, A. F. Harvey, "The Fundamentals of FFT-Based Signal Analysis and Measurement," Application Note 041, National Instruments Corporation, 2000.
5. M. Mayer, J. Schwizer, "Wire Bonder Ultrasonic System Calibration Using Integrated Stress Sensor," Semicon Singapore 2002, Advanced Packaging Technologies II, SEMI Singapore, pp. 169–175, 2002.
6. J. Schwizer, Q. Füglistaller, M. Mayer, M. Althaus, O. Brand, H. Baltes, "MEMS System with Multiplexer for In Situ and Real-time Wire Bonding Diagnosis," Advanced Packaging Technologies I, SEMI Singapore, pp. 163–167, 2002.
7. K. F. Graff, "Wave Motion in Elastic Solids," Dover Publications Inc., New York, 1975.
8. H. Fritzsche, "Improvements in Monitoring Ultrasonic Wire Bonding Process by Simultaneously Monitoring Vibration Amplitude and Bonding Friction Using Lasers and Analytical Models," VTE: Aufbau- und Verbindungstechnik in der Elektronik, **14**, Issue 3, pp. 119–126, 2002.
9. S. Y. Kang, P. M. Williams, Y. C. Lee, "Modeling and Experimental Studies on Thermosonic Flip-Chip Bonding," IEEE Transactions on Components, Packaging, and Manufacturing Technology, B, **18**, No. 4, pp. 728–733, 1995.
10. V. I. Fabrikant, "Applications of Potential Theory in Mechanics," Kluwer Academic Publishers, Dordrecht, 1989.

Chapter 4

1. R. E. Beaty, R. C. Jaeger, J. C. Suhling, R. W. Johnson, R. D. Butler, "Evaluation of Piezoresistive Coefficient Variation in Silicon Stress Sensors Using a Four-point Bending Test Fixture," IEEE Transactions on Components, Hybrids, and Manufacturing Technology, **15**, No. 5, pp. 904–914, 1992.
2. S. F. Chu, "Piezoresistive Properties of Boron and Phosphorous Implanted Layers in Silicon," Ph.D. Thesis, Case Western Reserve University, 1978.
3. G. G. Harman, "Wire Bonding in Microelectronics: Materials, Processes, Reliability, & Yield," McGraw-Hill, 2nd Ed., New York, 1997.
4. M. Mayer, J. Schwizer, "Wire Bonder Ultrasonic System Calibration Using Integrated Stress Sensor", Proc. SEMI Technical Symposium, Advanced Packaging Technologies II, SEMI Singapore, pp. 169–175, 2002.

5.   SPT, Small Precision Tools, http://www.sptca.com/frames-products.htm.
6.   NIST Property Data Summaries, Elastic Moduli Data for Polycrystalline Ceramics, http://www.ceramics.nist.gov/srd/summary/EmodOxRf.htm.
7.   J. B. Wachtman, and D. G. Lam, "Young's Modulus of Various Refractory Materials as a Function of Temperature," Journal of the American Ceramic Society, **42**, No. 5, pp. 254–260, 1959.
8.   D. S. Gardner, P. A. Flinn, "Mechanical Stress as a Function of Temperature in Aluminum Films," IEEE Transactions on Electron Devices, **35**, No. 12, 1988.
9.   E. Suhir, "Stresses in Bi-Metal Thermostats," Journal of Applied Mechanics, **53**, pp. 657–660, 1986.

Chapter 5

1.   A. Felber, W. Nehls, "Method and apparatus for measuring the vibration amplitude on an energy transducer", U. S. patent application publication No. 5199630, April 6, 1993.
2.   P. Hess, A. Greber, M. Michler, N. Onda, "Method and Device for Measuring the Amplitude of a Freely Oscillating Capillary of a Wire Bonder", US patent application publication No. 2003/0159514A1, August 28, 2003.
3.   M. Mayer, M. Melzer, "Method for the Calibration of a Wire Bonder", US patent application publication No. 2003/0146267A1, August 7, 2003.
4.   M. Mayer, O. Paul, D. Bolliger, H. Baltes, "In-Situ Calibration of Wire Bonder Ultrasonic System using Integrated Microsensor", Proc. 2nd IEEE Electr. Packaging Technol. Conf. EPTC'98, Singapore, pp. 219–223, 1998.
5.   M. Mayer, J. Schwizer, "Ultrasound Bonding: Understanding How Process Parameters Determine the Strength of Au-Al Bonds," Proc. International Symposium on Microelectronics IMAPS, pp. 626–631, 2002.
6.   M. Mayer, J. Schwizer, "Method for the Calibration of a Wire Bonder", US patent application publication No. 2003/0080176A1, May 1, 2003.
7.   M. Mayer, "Microelectronic Bonding Process Monitoring by Integrated Sensors," Ph.D. Thesis, No. 13685, ETH Zurich, Zurich, 2000.
8.   M. Mayer, J. Schwizer, O. Paul, H. Baltes, "In-situ Ultrasonic Stress Measurement During Ball Bonding using Integrated Piezoresistive Microsensors," Proc. Intersociety Electron. Pack. Conf. (InterPACK99), pp. 973–978, 1999.
9.   J. Schwizer, M. Mayer, D. Bolliger, O. Paul, H. Baltes, "Thermosonic Ball Bonding: Friction Model Based on Integrated Microsensor Measurements," Proc. 25th IEEE/CPMT Intl. Electronics Manufacturing Technology Symposium IEMT, pp. 108–114, 1999.
10.   Y. R. Jeng, J. H. Horng, "A Microcontact Approach for Ultrasonic Wire Bonding in Microelectronics," Journal of Tribology, **123**, pp. 725–731, 2001.
11.   —, "ANSYS Theory Reference," ANSYS Release 6.0, http://www.ANSYS.com, 2002.

12. J. Lubliner, "Plasticity Theory," Macmillan Publishing Company, New York, 1990.

13. Z. Zhong, K. S. Goh, "Analysis and Experiments of Ball Deformation for Ultra-Fine-Pitch Wire Bonding," Journal of Electronics Manufacturing, **10**, No. 4, 2000.

14. W. Budweiser, "Untersuchung des Thermosonic Ballbondverfahrens," (in German), Ph.D. Thesis, Technical Univ. Berlin, Berlin,1993.

15. Y. Takahashi, M. Inoue, K. Inoue, "Numerical Analysis of Fine Lead Bonding, Effect of Pad Thickness on Interfacial Deformation," IEEE Transactions on Components and Packaging Technology, **22**, No. 2, 1999.

16. D. S. Liu, Y. C. Chao, C. H. Wang, "Study of Wire Bonding Looping Formation in the Electronic Packaging Process Using the Three-Dimensional Finite Element Method," Finite Elements in Alanlysis and Design, **40**, pp. 263–286, 2004.

17. D. Tabor, "The Hardness of Metals," Clarendon Press, Oxford, 1951.

18. D. Stone, S. Ruoff, W. LaFontaine, H. Wilson, S. P. Hannula, C. Y. Li, "The Use of Indentation Techniques for Characterizing Wire Bonding Materials," 36th Electronic Components Conference Proc., pp. 318–323, 1986.

19. I. Kovács, G. Vörös, "On the Mathematical Description of the Tensile Stress-Strain Curves of Polycrystalline Face Centred Cubic Metals," International Journal of Plasticity, **12**, No. 1, pp. 35–43, 1996.

20. J. Seuntjens, "Gold Bonding Wire Alloys," http://www.kns.com.

21. E. Grüneisen, Ann. Phys., **25**, No. 4, pp. 825, 1908.

22. K. Hausmann, "Kurzzeitkristallisation und mechanische Eigenschaften von Feinstdrähten aus Gold und Kupfer," (in German), Ph.D. Thesis, No. 702, EPF Lausanne, Lausanne, 1987.

23. S. P. Hannula, J. Wanagel, C. Y. Li, "Evaluation of Mechanical Properties of Thin Wires for Electrical Interconnections," IEEE Transactions on Compnents, Hybrids, and Manufacturing Technology, **6**, No. 4, pp. 494–502, 1983.

24. F. Blaha, B. Langenecker, "Elongation of Zinc Monocrystals Under Ultrasonic Action," Die Naturwissenschaften, **42**, No. 20, pp. 556, 1955.

25. J. Herbertz, "Untersuchung über die plastische Verformung von Metallan unter Einwirkung von Ultraschall," (in German), habilitation, Gerhard Mercator Univ. Duisburg, Duisburg, 1979.

26. K. D. Lang, F. Osterwald, B. Schilde, H. Reichl, "Measurement of Ultrasonic Behavior During Wire Bonding - a Contribution to Quality Assurance in Chip on Board Technology," Proceedings Semicon West '98, pp. F1–F9, 1998.

27. J. Mattmüller, "Model for Ball Deformation During Ultrasonic Wire Bonding," Diploma Thesis, Physical Electronics Laboratory, ETH Zurich, 2002.

28. S. W. Or, H. L. W. Chan, V. C. Lo, C. W. Yuen, "Sensors for Automatic Process Control of Wire Bonding," Proc. 10th IEEE International Symposium on Applications of Ferroelectrics, **2**, pp. 991–994, 1996.

29. A. Carrass, V. P. Jaecklin, "Analytical Methods to Characterize the Intercon-
     nection Quality of Gold Ball Bonds," 2nd European Conference on Elec-
     tronic Packaging Technology (EuPac '96), DVS Berichte, **173**, pp. 135–139,
     1996.
30. S. W. Or, H. L. W. Chan, V. C. Lo, C. W. Yuen, "Ultrasonic Wire-Bond
     Quality Monitoring Using Piezoelectric Sensor," Sensors and Actuators,
     A65, pp. 69–75, 1998.
31. J. H. Cusick, A. E. Brown, A. S. Hamamoto, J. L. S. Bellin, "Ultrasonic
     Bond Monitor," US patent application publication No. 3890831, 1975.
32. J. Medding, M. Mayer, "In Situ Ball Bond Shear Measurement Using Wire
     Bonder Bondhead," Proc. 25th IEEE/CPMT Intl. Electronics Manufacturing
     Technology Symposium IEMT, SEMICON West, 2003.
33. J. Schwizer, M. Mayer, O. Brand, H. Baltes, "In Situ Ultrasonic Stress
     Microsensor for Second Bond Characterization," Proc. International Sympo-
     sium on Microelectronics IMAPS, pp. 338–343, 2001.
34. P. Palaniappan, D. F. Baldwin, "In Process Stress Analysis of Flip-Chip
     Assemblies During Underfill Cure," Microelectronics Reliability, **40**, pp.
     1181–1190, 2000.
35. J. C. Suhling, R. W. Johnson, A. K. M. Mian, K. Rahim, Y. Zou, S. Ragam,
     M. Palmer, C. D. Ellis, R. C. Jaeger, "Measurement of Backside Flip Chip
     Die Stresses Using Piezoresistive Test Die," Proc. International Society for
     Optical Engineering, pp. 298–303, 1999.
36. J. Schwizer, W. H. Song, M. Mayer, O. Brand, H. Baltes, "Packaging Test
     Chip for Flip-Chip and Wire Bonding Process Characterization," Proc.
     Transducers '03, pp. 440–443, 2003.
37. S. Wiese, "Experimentelle Intersuchungen an SnPb37 Flip-Chip-Lotkontak-
     ten zur Bestimmung werkstoffmechanischer Modelle für die FEM-Simula-
     tion," (in German), VDI, Düsseldorf, 2000.
38. P. M. Hall, "Forces, Moments, and Displacements During Thermal Chamber
     Cycling of Leadless Ceramic Chip Carriers Soldered to Printed Boards,"
     IEEE Transactions on Components, Hybrids, and Manufacturing Technol-
     ogy, **7**, No. 4, pp. 314–326, 1984.
39. J. H. Lau, "Solder Joint Reliability of BGA, CSP, Flip Chip, and Fine Pitch
     SMT Assemblies," McGraw-Hill, pp. 128, 1997.
40. L. K. Cheah, Y. M. Tan, J. Wei, C. K. Wong, "Gold to Gold Thermosonic
     Flip-Chip Bonding," Proc. 2001 HD International Conference on High Den-
     sity Interconnect and Systems Packaging, **4428**, pp. 165–170, 2001.
41. K. F. Reinhart, M. Illing, "Automotive Sensor Market," Sensors Update, **12**,
     pp. 213–230, Wiley-VCH, Weinheim, 2003.
42. J. Bernstein, "An Overview of MEMS Inertial Sensing Technology," Sen-
     sors, **20**, No. 2, pp. 14–21, 2003.
43. H. H. Bau, N. F. deRooij, B. Kloeck, "Mechanical Sensors," Sensors, **7**,
     VCH, Weinheim, 1994.

44.   N. Maluf, "An Introduction to Microelectromechanical Systems Engineering," Artech House, Boston, 2000.
45.   Y. Saitoh, K. Kato, M. Shinogi, "Semiconductor Acceleration Sensor," US patent application publication No. 6158283, December 12, 2000.

Subject Index